Measure of Emptiness

Creating the North American Landscape

Project Editor George F. Thompson
Consulting Editors Gregory Conniff, Bonnie Loyd, Edward K. Muller, David Schuyler

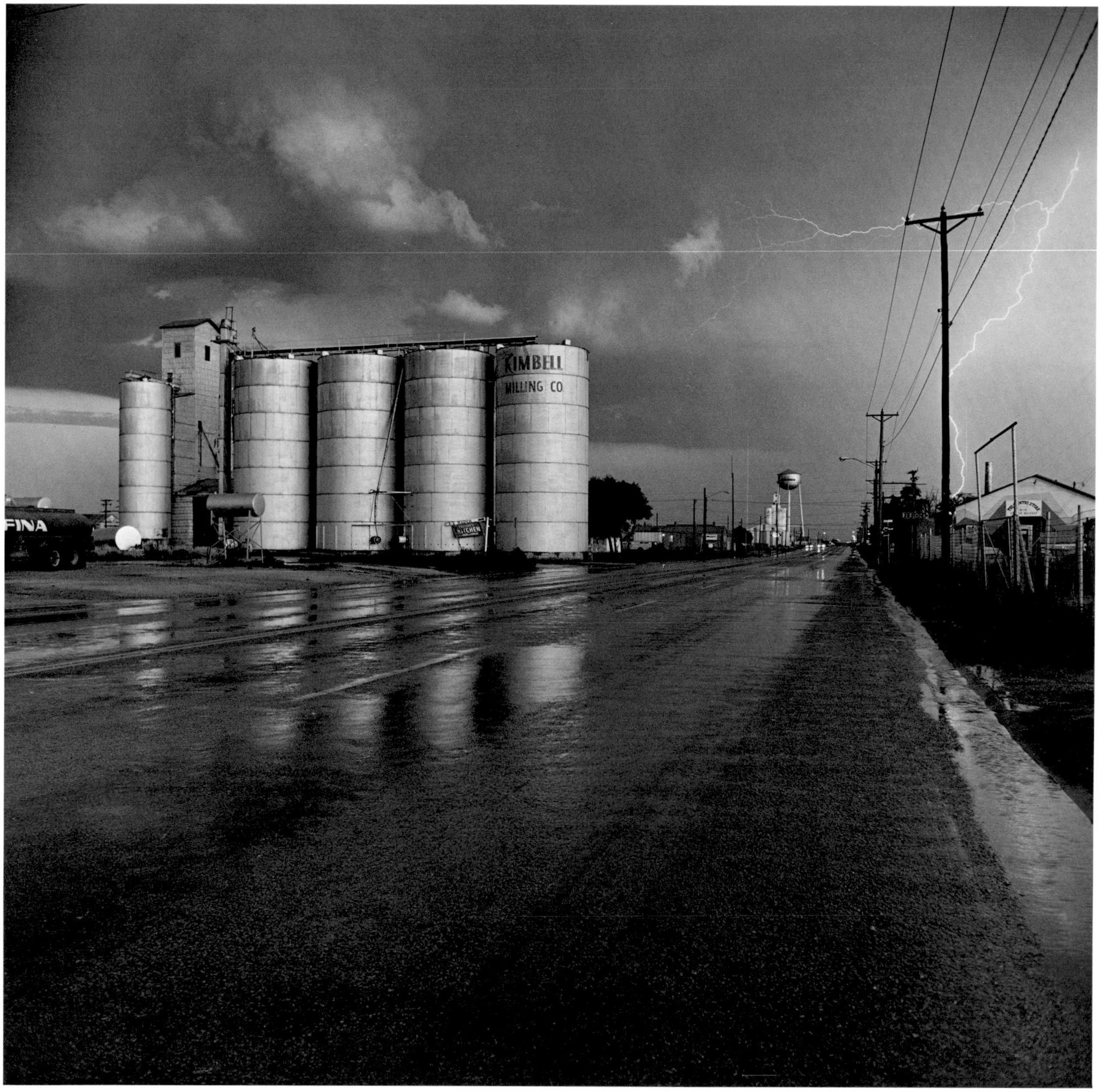
KIMBELL
MILLING CO.
FINA

Measure of Emptiness

Grain Elevators in the American Landscape

Frank Gohlke

with a concluding essay by John C. Hudson

The Johns Hopkins University Press / Baltimore and London

This book was published in cooperation with
the Center for American Places, Harrisonburg, Virginia.

Printed in the United States of America
on acid-free paper.

The Johns Hopkins University Press
701 West 40th Street, Baltimore, Maryland 21211-2190
The Johns Hopkins Press Ltd., London

Library of Congress Cataloging-in-Publication Data
Gohlke, Frank, 1942–
Measure of emptiness : grain elevators in the
American landscape / Frank Gohlke ; with
a concluding essay by John C. Hudson.
p. cm. — (Creating the North American landscape)
ISBN 0-8018-3928-9.— ISBN 0-8018-3949-1 (pbk.)
1. Grain elevators—United States—Pictorial works.
2. Landscape photography—United States.
I. Hudson, John C. II. Title. III. Series.
TR739.6.G64 1992
779'.3673—dc20
91-42844 CIP

(frontispiece)
Grain elevators and lightning flash / Lamesa, Texas 1975

(cover)
Grain elevator and plowed field / Wellington, Kansas 1973

For my mother, Nellie May Ross Gohlke,
and in memory of my father, Robert Lee Gohlke

Contents

Acknowledgments

Many people and organizations have helped to make this book a reality. The Minnesota State Arts Board, the John Simon Guggenheim Memorial Foundation, and R. Hugh Uhlmann of Standard Milling Co., Kansas City, Missouri, provided crucial financial support during the time I was making the photographs. David Travis of the Art Institute of Chicago and John Szarkowski of the Museum of Modern Art, New York City, offered encouragement through exhibitions of the photographs. Kent and Marcia Minichiello and Charles Melcher contributed generously to the publication phase of the project.

I have had the good fortune to have many fine teachers, most of whom are also friends and colleagues. Paul Caponigro was my first example of what it means to live and work from the inside out. J.B. Jackson taught me how to interrogate a landscape. Karal Ann Marling and the late Reyner Banham were fellow enthusiasts for grain elevators. Robert Adams, Tom Arndt, Bob Bergman, Gregory Conniff, Stuart Klipper, Ben Lifson, and Nick and Bebe Nixon have in their several and unique ways deepened for me the connections between living and photographing. To my daughter, Jessica Gohlke, go special thanks for saying to me out of the blue at age nine that of all my pictures she liked the grain elevators the best. An opinion like that keeps one going for a long time.

The grain elevator project might never have come to its fruition in this book had not Gregory Conniff introduced me to its editor, George F. Thompson, whose understanding of the project was immediate and whose advocacy never faltered. Peter Galassi's and John Fraser Hart's suggestions helped me to produce a stronger essay. Thanks to Lorraine Ferguson, Robert Hennessey, and the staff at the Johns Hopkins University Press for their craftsmanship and professionalism.

And I am deeply grateful to my wife, Lucy Flint-Gohlke. Her sensitive eye, her unerring ear, her alertness to the large effects of small details, and above all, her unstinting gifts of time and attention, are manifest throughout this book.

F.G.

Measure of Emptiness

Measure of Emptiness

by Frank Gohlke

In the autumn of 1971, after seven years in New England, I moved with my family to the Midwest. The windows of our hilltop apartment gave onto a comprehensive view of the Midway, a mile-long stretch of grain elevators and railroad tracks on the boundary between the cities of Minneapolis and St. Paul. During the months of disorientation following the move, I would often stare idly out the window, content, after my experience of the intimate, crowded landscapes of the eastern seaboard, to be once again in a place where my eyes could find real distances. As I gained my bearings, the grain elevators of the Midway began to draw my attention. Their scale, featureless surfaces, and simple repetitive forms gained a hold on my imagination that I could not fully comprehend.

Grain elevators were prominent buildings in my hometown of Wichita Falls, Texas, but the terminal elevators of the Midway were different—bigger and more numerous than the ones I was used to. Seen end-on, as they were from the Kasota Avenue overpass on Highway 280, they were like a city: their proportions were similar to those of typical office buildings, while the rhythms of their massing mimicked those of the downtown Minneapolis skyline two miles to the west. The analogy with the modern city broke down, however, when the elevators were viewed at closer range. On my frequent detours across the tracks and along the rutted dirt roads that ran between the elevators, I found an area that was rough, unfinished, and empty. University Avenue, a major thoroughfare connecting Minneapolis and St. Paul, was only a few blocks away, but it could have been miles, or centuries.

At first I just savored the strong emotions the place provoked, which mixed the awe one feels in the presence of monumental architecture with the impatient curiosity of an archaeologist at a new site. The place encouraged fantasies of lost worlds and vanished empires, of abandoned cities whose makers' intentions were utterly inscrutable to me. The grain elevators' resemblance to habitable architecture, however, only serves to dramatize the differences. The windowless, largely unbroken expanses of concrete or corrugated steel ten or more stories tall and hundreds of feet long produce a strange sense of dislocation when one is near them. The curved sides of the concrete elevators do unexpected things to shadows, transforming straight lines into arcs, drooping wires back into straight lines. The sounds that come from inside the elevators—whining, thrumming, generic machine noise—reverberate in the deep alleys between rows of bins. The light seems to come from far away. Other people are seldom seen.

The physical power of the Midway is undeniable, but that power alone cannot account for the strength of my attraction to the place. I began photographing grain elevators at a time when I was fascinated with blank walls, empty facades, shuttered windows, and all manner of unyielding surfaces, emblems of an interior crisis. It is also true, and just as important, that the grain elevators represented a connection to my childhood home, toward which I was feeling a renewed sense of curiosity and attachment. It is equally true that I sensed in the grain elevators a richer set of relationships with a broader world than was offered by the other subjects with which I was working. I was beginning to understand that my photographs interested me more when they were open to issues beyond aesthetic ones.

Many of my earliest grain elevator photographs were close-ups, in the sense that the elevator and its attendant structures would fill the frame. The fact that these "close-ups" were often made fifty to one hundred feet away from the subject gave me my first insights into the difficulty of accurately describing physical scale in a photograph. These pictures when printed appeared essentially scaleless, and yet my strongest physical sensations when taking the picture had to do with feeling dwarfed by the elevators' size. The discrepancy stemmed in part from thinking about the photograph as something that took place entirely within the frame, which I emphasized by printing the black of the clear film around the image. This was a sign of the purity of my commitment to what the camera saw, but it also diluted the connection between the pictorial space and the three-dimensional space from which it was derived. At the outset of the project I did not see this as a problem; I thought of the Midway as little more than a mine with an inexhaustible vein of good-looking pictures that I could extract until I lost interest in them. When I began to wonder, however, about the place itself, about what created its character and how the photographs might reflect that, the issues changed. The pleasure I took in the forms and surfaces of the structures that comprised the Midway was augmented by curiosity, which, as it often does in life, led me in unforeseen directions.

There is literally a great deal more to a grain elevator than meets the eye. I began asking questions of the workers I encountered on my visits to the Midway. What is stored inside, and what happens to it while it's there? How are the elevators constructed? What are those big, funnel-like structures hanging on the sides? What is your job? The answers were delivered in slow, flat, Minnesota English by polite men who were as bemused by my interest in them and their work as I was interested in the information they were giving me. I kept hoping that one of my informants would open a door to the larger dimensions which I was sure the elevators inhabited, but their responses were insistently straightforward and factual. The features of the elevators that I found so visually compelling all had predictably mundane explanations, and the people who had daily contact with them had, also predictably, little inclination to see them in other terms. My enthusiasm for a particularly beautiful surface was received with ironic amusement by a foreman in green coveralls, to whom it was an eyesore, caused by a bad batch of black paint that had weathered quickly to reveal the swirling marks of the concrete workers' trowels.

People who work in grain elevators spend most of their time tending the machinery: conveyor belts, the vertical elevators (known as "legs"), movable spouts, automatic hoppers, fumigators, scales, and fans. Other workers grade the grain and test its moisture content, keep track of the grain coming in and going out, and decide in which bins it will be stored. Grain usually does not sit still for long inside a grain elevator, no matter how long it remains there. It has to be cleaned of dust (which is collected in the funnel-shaped cyclones on the exterior) and other foreign matter, treated with fungicides and insecticides, and periodically aired out, all done on schedules that vary according to the type and quality of the grain. Work inside the elevators is carried on mostly in the galleries above and below the storage bins, where the grain is always in motion, and so the workers are exposed to a great deal of dust. Besides being highly explosive, grain dust can cause respiratory illness, so prudent workers protect their lungs by wearing disposable dust masks, which cover the nose and mouth. The masks lend them an alien aspect, particularly when they are seen surrounded by the ochre clouds of dust that regularly issue from the doors.

As interested as I was in what I was learning, I was frustrated by the discrepancy between the ordinariness of the facts surrounding the grain elevators and the intensity of my emotional response to the objects themselves. What was the source of the excitement I felt while photographing the elevators of the Midway? Why did my heart race as the prints came up in the developer? True to my academic training, I thought I might find the solution to this conundrum in books. My initial reading did not yield much that helped me. Most of the immediately relevant material was very technical, but from such volumes as *Scientific Principles of Grain Storage* and *The Design of Walls, Bins, and Grain Elevators* I did learn that grain can be stored satisfactorily in any number of ways, including underground. This supported my hunch that the form of grain storage that had evolved in America might be the result of more than an inescapable engineering logic applied to the large-scale production and distribution of grain. To me, the photographs I was making argued that there are deeper impulses lurking somewhere in the functional surfaces and details of the grain elevators, and that subjective choice as well as objective necessity has a role in determining their form.

The rationalist in me, however, was leery of drawing large conclusions from a single kind of evidence, no matter how persuasive, and I was acutely aware as well that the evidence offered by photographs says as much about how a thing is perceived as about what it is. The photographs gave eloquent testimony to the poetic possibilities of grain elevators. They were mute, however, on the question of whether that poetry was audible to anyone but me. I felt sure that there were aesthetic and symbolic dimensions one could not ignore if one wanted to understand the grain elevators fully, and that these intangible qualities arose from concrete relationships among the buildings, the people, and the land. Were my photographs tuning in to a level of intention in the grain elevators hidden perhaps even to their makers, or were they purely projections of my own need to see the subject in a certain way? I thought that if I could discover others who shared my intuitions or could give them a foundation in fact or theory, then I could proceed more confidently.

Several people directed me to the work of the French architect and theorist Le Corbusier, who was so impressed by photographs he saw of North and South American grain elevators that he devoted a chapter to them in his influential polemic, *Vers une Architecture* (1923). Although it was heartening to discover that I was not alone in my conviction that grain elevators were buildings with important lessons to teach, it was immediately clear to me that Le Corbusier was interested in them for reasons that were very different from my own. Only one of the illustrations in *Vers une Architecture* depicted an elevator located in the central plains, a setting that was quickly becoming as important to me as the buildings themselves. I had better luck when I found two tantalizing statements in a large compendium of technical and historical articles, advertising, and Chamber of Commerce publicity, called *Grain Elevators of North America* 5 (1942):

> While designed primarily to get the greatest capacity for the least money, the new elevator at Ogden, Utah, like plants of the Sperry Flour Co. in other cities, shows a proper appreciation of artistic appearance. (217)
>
> Elevators have always been constructed with an eye to utility rather than art. But by the judicious use of black paint, brushed heavily on cornices and around the foundation line, Chalmers and Benton have added a touch of artistry to the toneless, dull concrete that makes the building stand out attractively against a western sky.
> ("Star Elevator at Bison, Oklahoma," 338)

"Against a western sky." The suggestiveness of that phrase was soon given substance when, in the early spring of 1973, I drove with a friend from Minneapolis to Albuquerque, New Mexico, to attend a conference. Our route took us into the southwestern corner of the state of Kansas, which advertises itself as, among other things, "the Wheat State." We traveled on U.S. 56 from Great Bend through Dodge City to Sublette, where we turned south to Liberal, on the Kansas-Oklahoma border, and joined U.S. 54 heading for Tucumcari, New Mexico. Along the way we passed grain elevator after grain elevator, one every ten miles or so. Most of them were painted white, and all of them dominated the towns and surrounding landscapes with the irrefutable authority of a hilltop fortress in Tuscany or a colonial church on a New England common. The trip was a revelation; it altered my conception of the project in fundamental ways, moving its center of gravity from the city to the country, from a pictorial space bounded by the frame to an unbounded space to which the frame served as entrance.

Certain things did not change, particularly the questions provoked by the grain elevators wherever they were located, questions about their builders' intentions, questions about what else the buildings might represent beyond, in Reyner Banham's words, "an almost excessively good investment." But a new element had been added that refocused these speculations: the landscapes of the prairies and High Plains. Now it was not just the buildings themselves that I wanted to account for, but the resonance I sensed between the structures and their locations along the highways in Kansas I had recently traveled. They illuminated one another; the grain elevators helped me to see the landscape, and the landscape caused new aspects of the grain elevators to stand out in relief.

It was as forms in light that I first responded to grain elevators. Their surfaces held and returned light in ways that looked wonderful in a photograph, and for a time the play of light and shadow in a shallow space provided ample scope for exploration. But light can do more than activate surfaces. Light creates and intensifies the photograph's illusion of volume and mass; and when I was standing on the edge of a small town in Kansas or Oklahoma, the mass of the grain elevator and the volume of the empty space around it seemed much more central to my enterprise than the beautiful patterns of shadow on a concrete surface. A grain elevator is not only massive; it is also tall, and the endless horizontals of the surrounding landscape underline its vertical thrust. Tall buildings amaze us in part because of the difference in scale between them and us, and in part because we understand intuitively the magnitude of the force that must be overcome in order for them to stand. The grain elevator's uniform surfaces and absence of windows make it look solid, heavier than other buildings, and thus its defiance of gravity seems more dramatic. The weight of the grain elevator is no illusion; its walls, built to contain the force of a full column of grain, are often two feet thick or more at the bottom. Its solidity, however, is only apparent. A grain elevator is in fact a container, a volume to be filled, a shell surrounding more empty space.

The relationship between the space inside the grain elevator and the space outside intrigued me. The locations of the early elevators had been dictated by the available means of transportation and the suitability of the terrain, soil, and climate for a particular kind of agriculture. Their size and number in a given area were measures of the land's productivity and the industry of its inhabitants. All of these factors could and did change, and when they did the fortunes of a particular grain elevator might change as well, for the better or for the worse. In order to account for what I was seeing, it seemed, nothing could be left out. The more deeply involved in the project I became, the harder it was to fix its boundaries, or even to say with certainty what "it" was.

I was clear, however, about one thing: the grain elevators could not be considered in isolation from the landscape; the building and its context were inseparable. At the same time, I was beginning to realize that the landscape is not a collection of fixed objects on a static spatial grid but a fluid and dynamic set of relationships. Its appearance is the result of a multitude of forces acting in time on the land itself and its human accretions.

The contact sheets from my first grain elevator expedition in the summer of 1973 suggest that I was already thinking in these larger terms, whether I was aware of it or not. A quarter of the frames have no grain elevators and often no buildings at all. In half of the frames the grain elevators are in the middle to far distance and take up a small fraction of the picture area. In only one-fourth of the pictures made on the trip do grain elevators dominate the frame in the way one expects a presumed subject to do. My conceptual and experiential understanding of landscape and my definition of the project were evolving together, both of them shaped by the view from the road.

What struck me first of all on this trip was how good I felt in places where the conventional wisdom about beauty and scenery told me I should be bored and uncomfortable. As a child I had visited many of the scenic wonders of the West with my family. On our way, we would have to drive through the Texas and Oklahoma panhandles, Kansas, eastern New Mexico, or eastern Colorado, and we would do it as quickly as possible, often at night, because such country was too similar to what we lived with every day. The real scenery was understood to lie beyond these unprepossessing lands, which seemed flat as table tops and, to a child, endless. Now as an adult I was drawn to those very places. The flatter it was, the better I liked it. Hills were disappointments, because hills meant something was hidden.

These sentiments were not peculiar to me alone. People I encountered on my grain elevator trips said things like, "The mountains are beautiful, but I couldn't live there for long. I need to be able to see five miles in any direction to feel right." In places where there is much to see, where visual incidents are packed so densely that the eye is inundated wherever it falls, my responses tend to become passive and conventional. In places where there seems to be nothing to see, my eyes have to work harder, and so, paradoxically, I see more. Beauty and grandeur can be distractions when one is trying to understand how a given piece of land has been shaped. The writer J. B. Jackson, an enthusiast of geography, alluded to this problem when he impatiently dismissed the national parks as "just scenery" having nothing to do with landscape. Landscape is an active principle. Its existence is the result of human actions and natural processes in ever-changing combinations, and its understanding requires that the senses, the mind, and the imagination be fully engaged with the facts in front of us. My study of landscape began when I asked, "Why do these things look the way they do?" and it was furthered when I discovered, through a chance encounter with a geographer at a dinner party, that a group of people had been posing that question in many forms for years.

I began to read the work of geographers Carl Sauer, David Lowenthal, and Yi-Fu Tuan, and the essays of J. B. Jackson. These writings confirmed my intuitions about the relationships between people and places, about the deep and continuing interchange between the given character of an environment and its inhabitants: what is built reflects both inner needs and external constraints and becomes in turn a part of the landscape that influences subsequent generations. Thus, a grain elevator may be at one and the same time a characteristic element of the landscape, an embodiment of a particular response to a specific set of circumstances, and a multivalent symbol to the community it serves and to outsiders. For me, however, simply knowing these things was not enough; as a photographer I had to identify those places where these abstract concepts took on a palpable form. The ideas were of no use to me unless they could be seen and felt.

I found what I was looking for in places that most people went out of their way to avoid; and it was Jackson, of all the writers I read, who most explicitly affirmed my belief in the beauty and meaning I was finding there. He strengthened my understanding of the responsibility of the landscape photographer: to make the invisible visible, to see clearly and unsentimentally what has heretofore escaped our attention. Above all, he and his colleagues articulated an obvious truth that struck me with the force of revelation: *everything* in a landscape has meaning.

I had not held the opposite to be true; I had simply never put it to myself that way before. And it is not just that each thing has its own discrete significance, but that every element of a landscape participates in many networks of meaning, so that asking a simple question about a common object can open up the human history of an entire region. Jackson can begin with a simple description of different types of fences, and before long we find ourselves considering changing attitudes toward property and privacy, the evolution of animal husbandry in response to increasing population density, and the influence of nostalgia on our judgments of landscape beauty. To see the landscape as a tissue of relationships is a natural consequence of the assumption that landscape is a human creation, even when the only action involved is the act of perception. Space is given its character, place comes into being, by the way human presence is manifested within it. At the same time that I was assimilating these ideas, I was becoming more interested in subjects that were less material, such as the way space is divided and shaped. It occurred to me that relationship itself could constitute a subject.

The photographs were changing, too. For one thing, by the time of my second grain elevator trip in the summer of 1975, I had become very scrupulous about not tilting the camera up or down. As I worked with the images from the 1973 trip, I became increasingly uncomfortable with the convergence of vertical lines that resulted when I tilted the camera up to take in the top of a structure. The convergence reminded one of the lens, and as the pictures had more and more to do with the particular qualities of space, that became an intolerable distraction. It seemed to me that if I could convince the viewer of the neutrality of the description, then there was a chance he or she might notice what had arrested me and made me want to take a photograph. Often the impetus was so close to the margins of perception that, when I looked at the contact sheets, I was uncertain if I could see it myself. For subjects that were so elusive, an obvious appeal to style was an obstacle to the apprehension of content.

There were other reasons as well for my reluctance to tilt the camera, reasons that had nothing to do with thought. Surrounded by the immense spaces of the Kansas prairie, it felt right to "compose" as little as possible, particularly on the vertical axis. I tried to make myself as vertical as I could and the camera as horizontal as possible, thus reproducing physically the prime orientations of that landscape. The visual response to the depiction of the buildings and the physical response to the landscape reinforced each other; the eye, the mind, and the body all contributed to a decision that in turn influenced the picture's content in ways I hadn't anticipated. Holding the camera level while standing in a flat landscape means that the horizon cuts the middle of the picture and divides it into equal portions of earth and sky. I gradually became aware that the bisected frame was for me emblematic of that landscape, with sky and earth in a dynamic equilibrium and human beings in the middle, exploiting both and being dominated by both. Thus, a pictorial element that followed from other considerations became in itself a thematically significant aspect of the project. Conscious programs, deep convictions, and unexpected intuitions all influenced the making of the photographs, and the photographs contributed to my understanding of all three.

To design and build a safe, economical structure for the storage of grain requires knowledge of structural and mechanical engineering, fluid dynamics, the control of volatile gases, and many other subcategories of applied chemistry and physics, to say nothing of the business acumen and agricultural expertise necessary to predict whether such a venture will succeed financially. It might seem that there would be little room in these calculations for such essentially poetic concerns as a feeling for the local landscape, but this is to forget that our actions arise from many sources at once, some of them unacknowledged, and that response to place is a fundamental component of all our experience. The structures that are shaped by this response embody our values; our choices about what to build and how to build it reveal what is important to us. Grain elevators are the product of a culture that favors the sense of sight above others. We need to *see* our presence manifested in the landscape, preferably over great distances. The workaday operations of the grain elevators could cease, and they would still serve a vital function on the prairies simply by being so definitively *visible*. The centrality of these buildings, however, is not limited to the practical and the perceptual; they summarize persistent antinomies in American civilization.

When Le Corbusier called grain elevators the "cathedrals of the prairies," he was saying more than he knew. A woman in Plainview, Texas, told me, "Out here the churches don't need to have tall steeples—we have the grain elevators," confirming Le Corbusier's intuition and highlighting the amalgam—some would say confusion—of the spiritual and the material that is so characteristic of American public mythology. If grain elevators are secular cathedrals, is it because the spiritual has invaded the domain of the material, transforming dross into gold, or has the material appropriated the spiritual, reducing it to a hypocritical justification for the exploitation of the land and the pursuit of wealth? It is not necessary to portray the issue in such judgmental terms to recognize that the vertical reach of the grain elevator proposes a connection between human industry, the land's bounty, and divine favor. Whether that claim is advanced in a spirit of thankfulness or of self-congratulation is unclear, one of the ambiguities that makes these buildings such rich objects of contemplation.

For me, the essential grain elevator view is obtained through the windshield of a car or truck while traveling on a highway in Kansas or Oklahoma or the Texas Panhandle. It is not a static view, but one that begins just as the elevator becomes visible above the center line, about five miles out of town, and continues until it disappears in the vibration in the rearview mirror. In the minutes that pass as the speck grows to colossal size and then shrinks to rejoin the horizon, many contradictory messages are created: we are powerful, we build for the centuries, our monuments rival those of other heroic ages; we are insignificant, our hold on this landscape is tenuous, nature and time erode our greatest creations as if they were dust. What lingers in the memory, though, is the image of a solitary, upright form in the middle distance of an endless plain. Compare this to a Hopi village or a New Mexico pueblo, also built by human beings who inhabit places vast in scale, and it becomes an emblem, exaggerated almost to the point of parody, of the American glorification of the individual. But this visual association must coexist with the fact that, particularly in the small towns of the Midwest, the grain elevator is a center of social and economic life and, by its constant, far-flung visibility, symbolizes to the inhabitants the coherence, vitality, and continuity of their communities.

The previous sentence might well have been written in the past tense. A great deal has changed since I began the grain elevator project in 1972 and ended it in 1977, after a year of photographing mainly elevators that had exploded, burned, or were being demolished. I thought at the time that these images of destruction only signaled to me that I needed to find a new subject, but perhaps I was making explicit an element of the project that I had previously chosen not to notice. When I began photographing in the Minneapolis Midway, it was already an area in decline. Some of the elevators were only partially used; some were empty and, as I was to learn, slated for demolition. A large malting plant was shut down and its elevators leased for storage. The older elevators were in bad repair, and many of the newer ones needed paint.

The deteriorating fortunes of the Midway signaled changes in the marketing and storage of grain that accelerated during the 1970s. When these structural shifts combined with the farmbelt depression of the 1980s, many grain elevators, both in the country and in the city, became obsolete, regardless of the fact that their solid construction may enable many of them to outlast the culture that produced them. What had seemed at the outset to be a celebration of a key and enduring feature of the prairie landscape now appears to be an elegy.

"In the United States there is more space where nobody is than where anybody is. That is what makes America what it is." When I happened on Gertrude Stein's dictum in 1974, it helped me to identify what I wanted to photograph: the "space where nobody is," the emptiness of the prairies. Grain elevators were the presence against which that emptiness could be measured, the material response of a group of people to a vastness for which their history and traditions gave them little preparation. When I structured this book in 1978, I removed the grain elevators from the stage at the end in order to get a clearer view of the set. The intervening decade has given that decision an unintentionally prophetic and ironic turn, but topicality was not then the point, nor is it now. The elegiac note is appropriate regardless of the ultimate fate of the grain elevator, because these monumental structures remind us, as do the great human achievements of every age, of what passes and what endures.

Ashland, Massachusetts 1991

PLATES

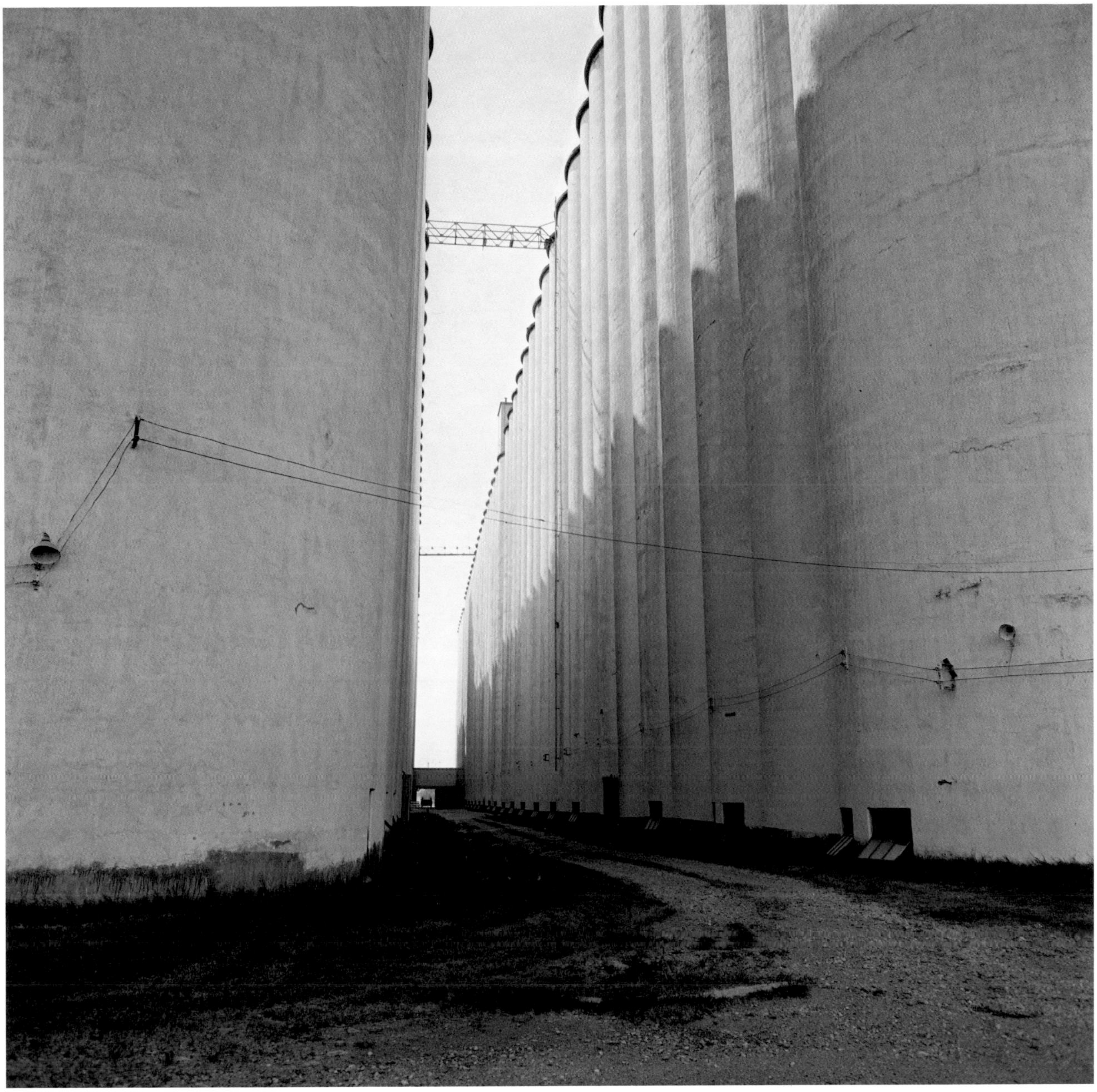

Large grain elevators / Enid, Oklahoma 1973

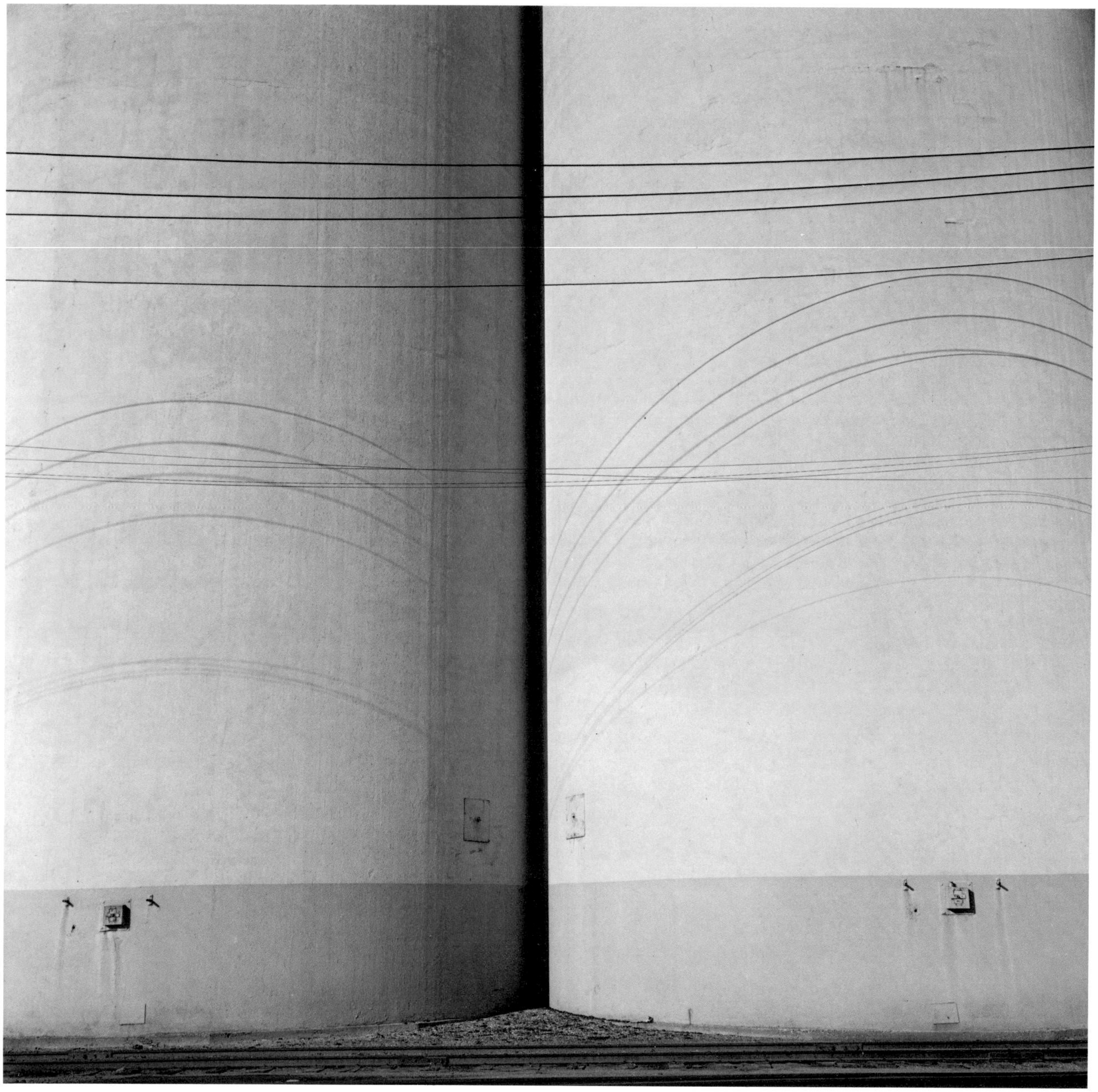

Grain elevator / Midway area, Minneapolis, Minnesota 1972

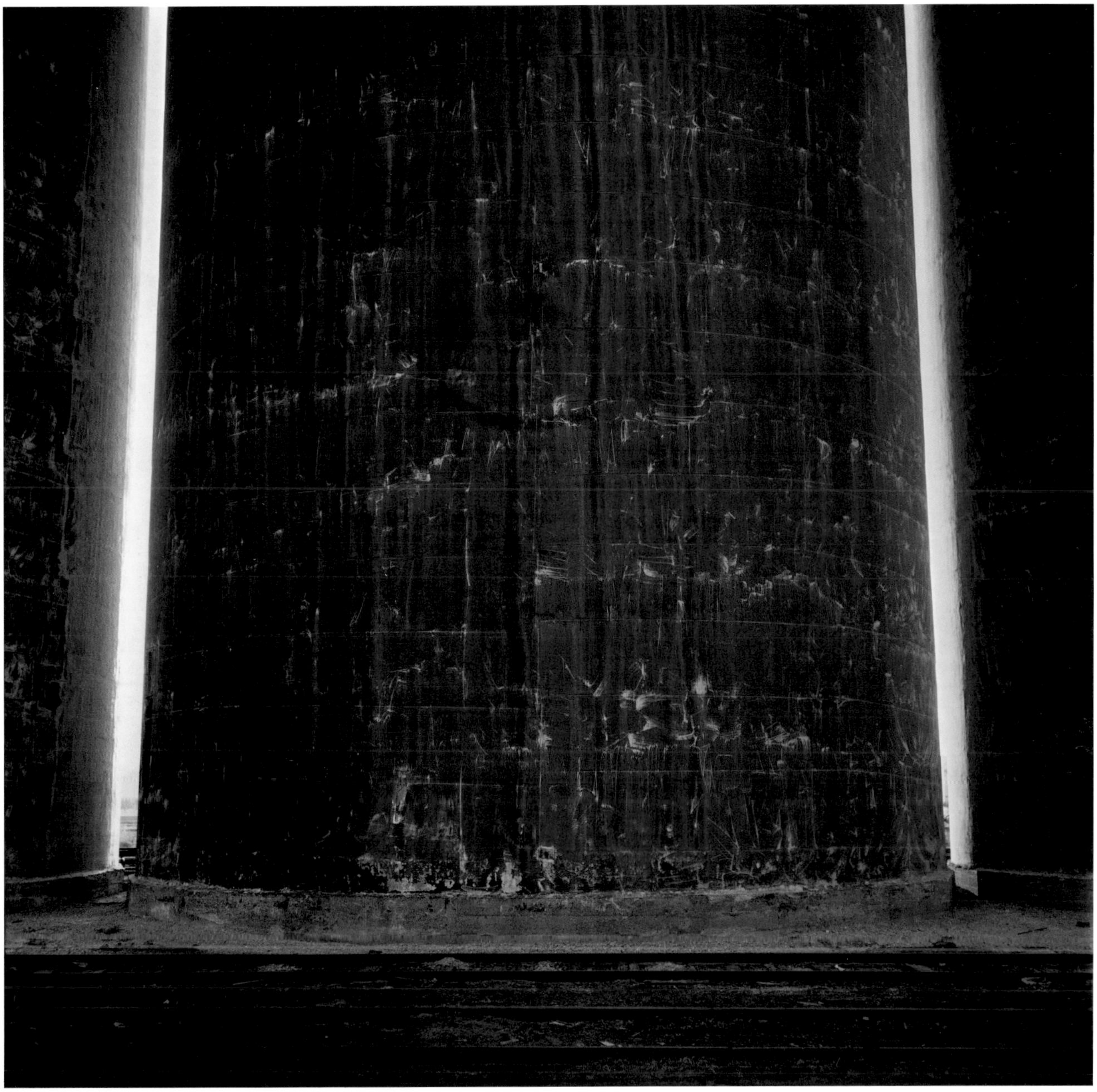

Grain elevator / Midway area, Minneapolis, Minnesota 1972

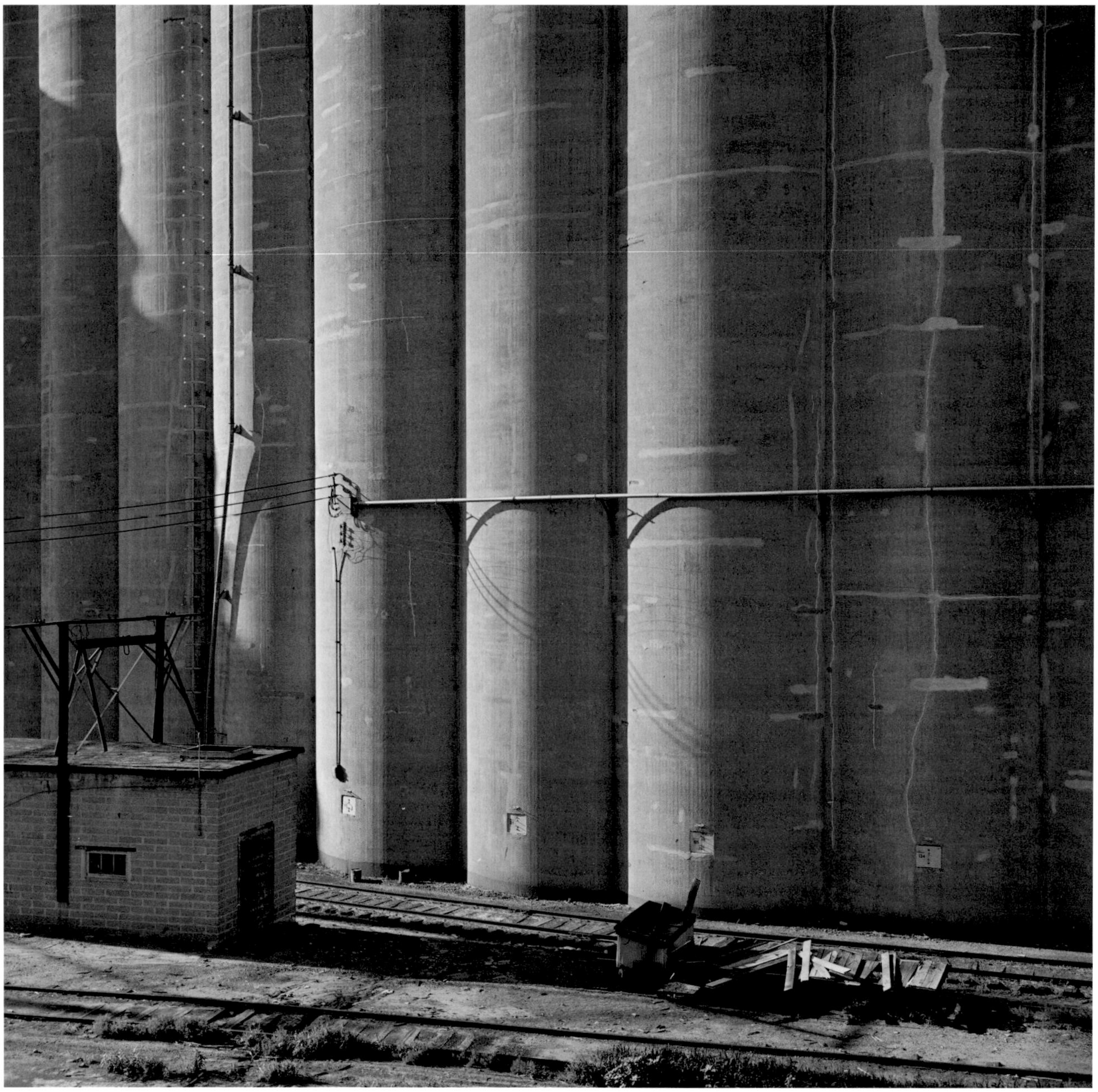

ADM grain elevator / Midway area, Minneapolis, Minnesota 1973

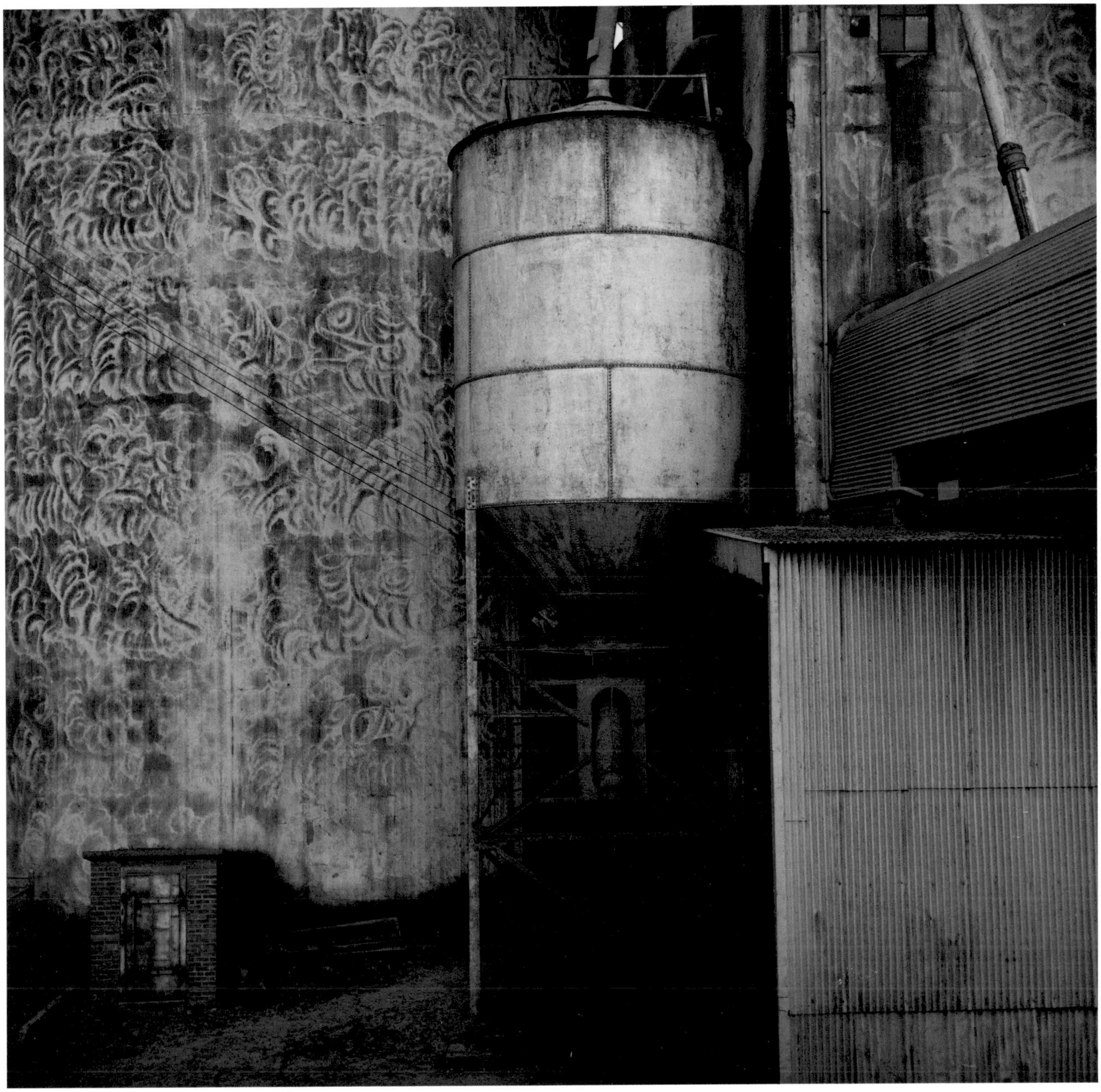

Bins, hopper, and loading shed / Midway area, Minneapolis, Minnesota 1972

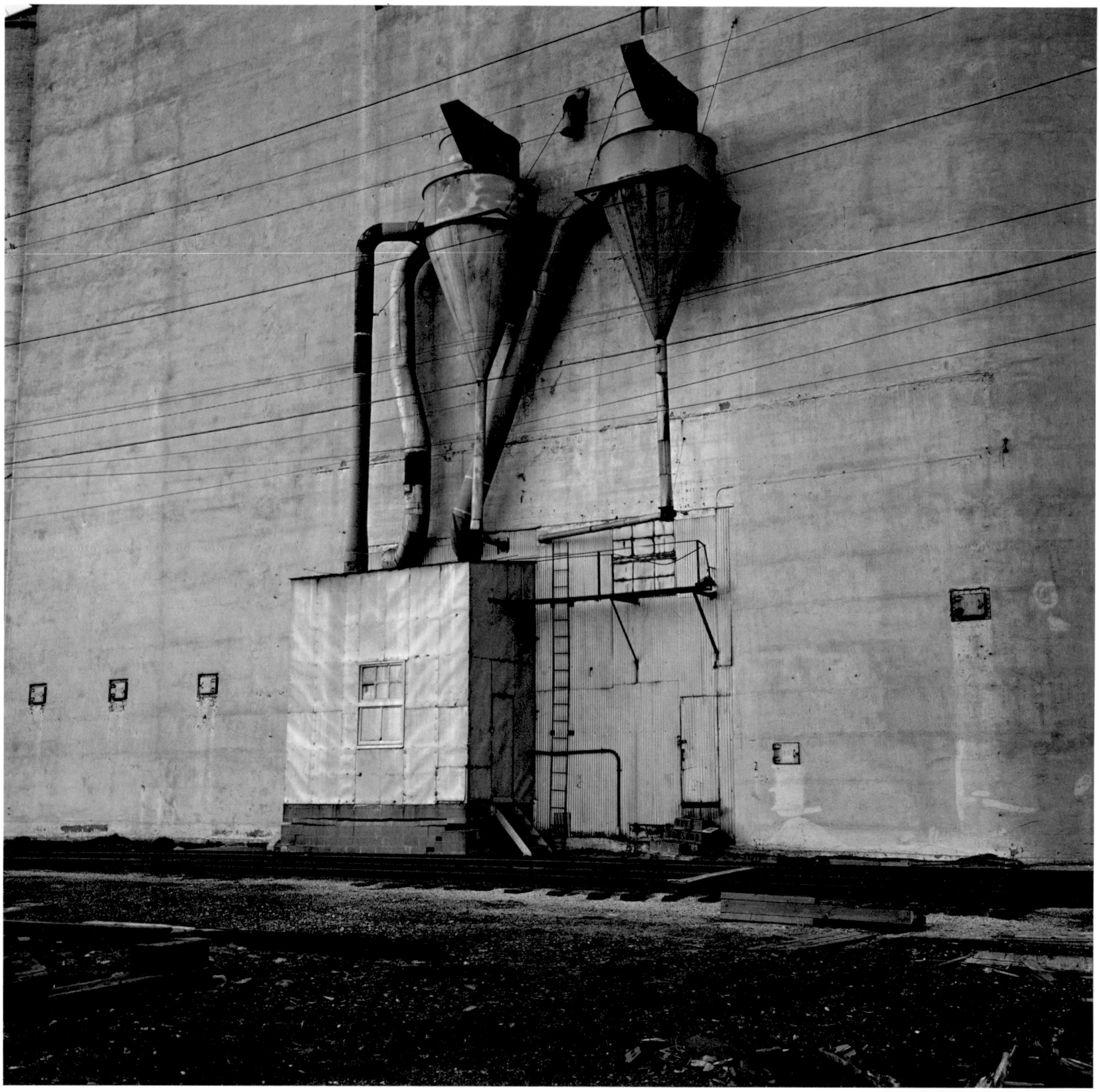

Cyclones (dust precipitators) / Midway area, Minneapolis, Minnesota 1973

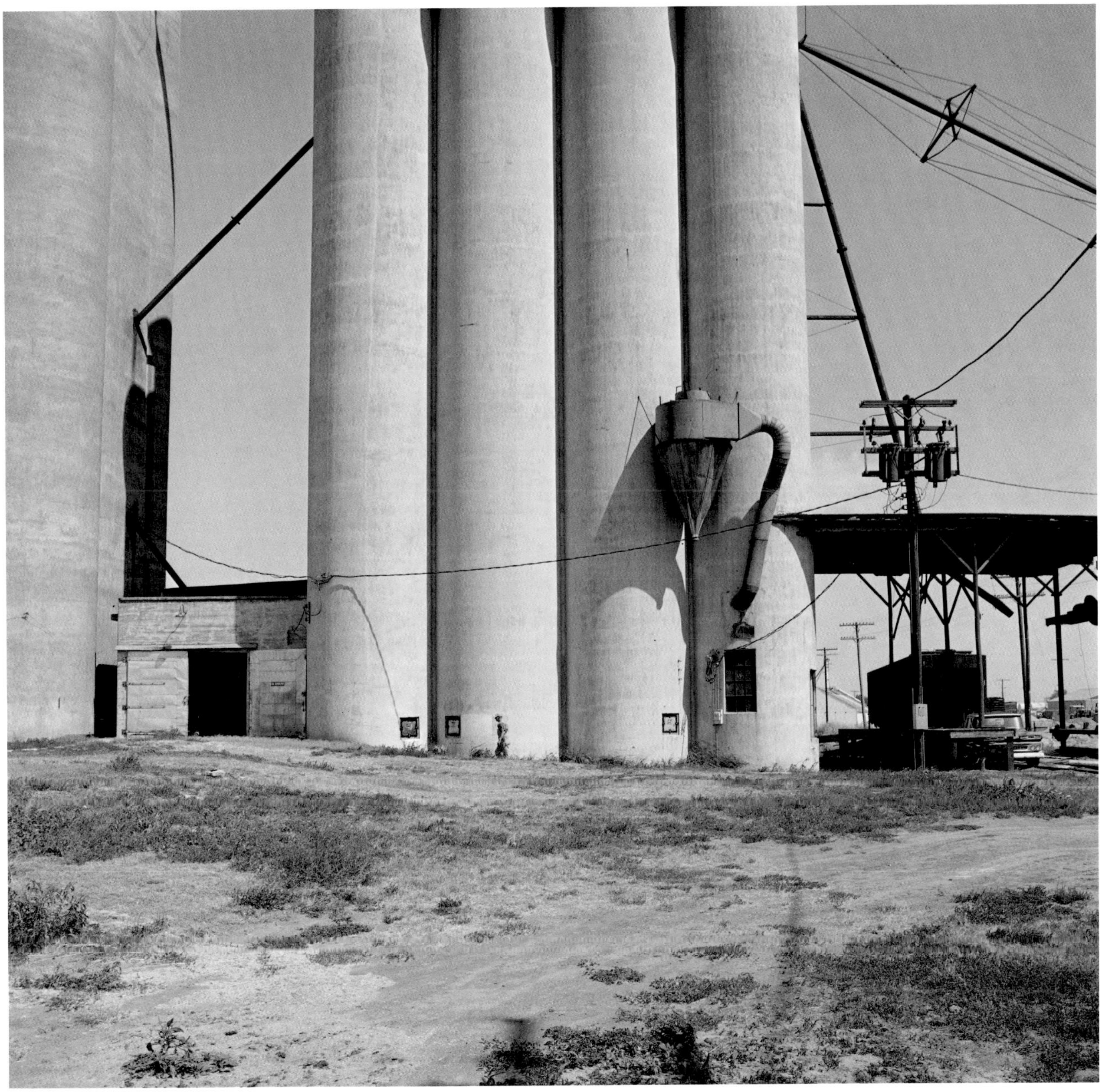

Grain elevators / Hennessey, Oklahoma 1973

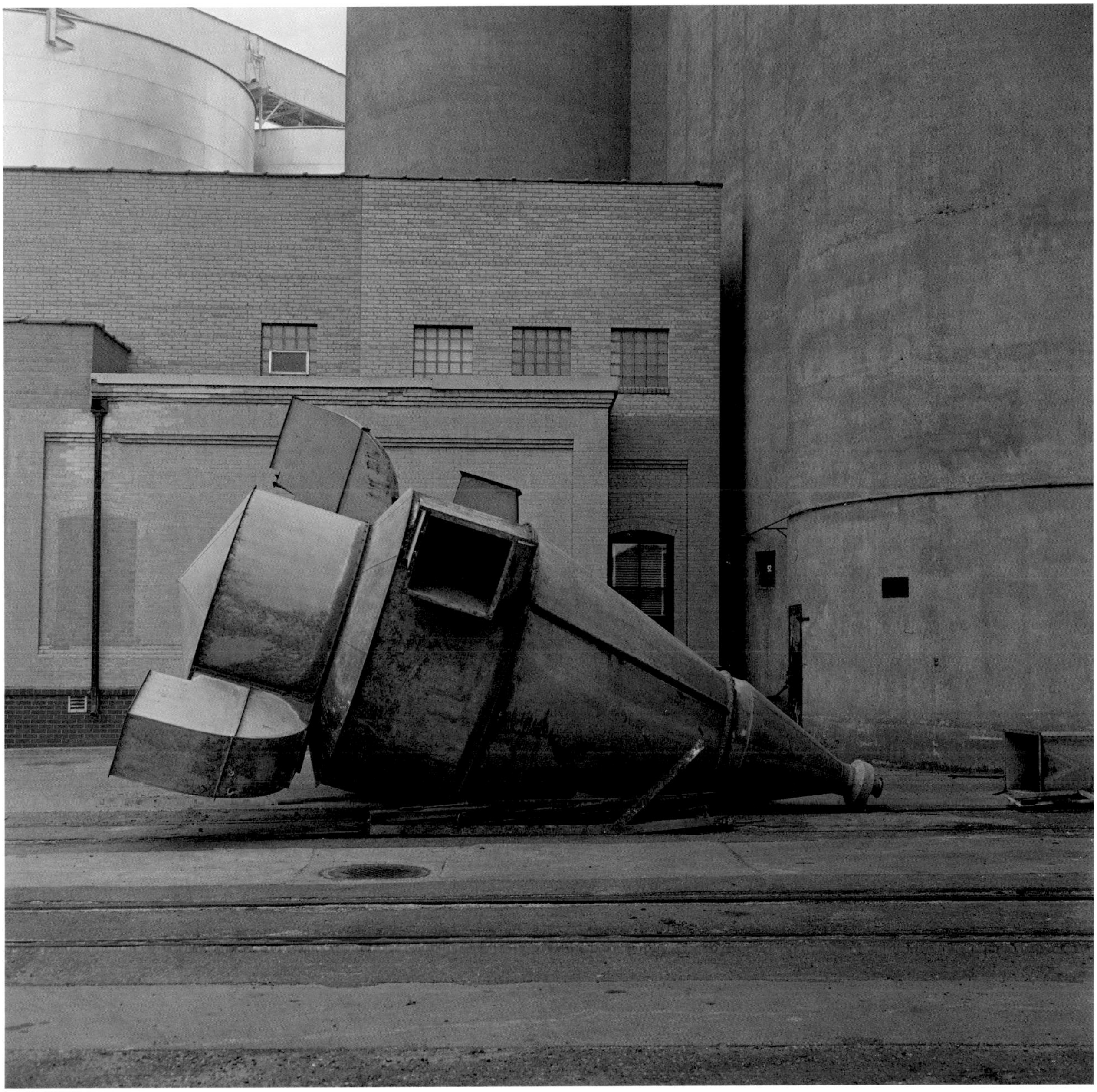

Cyclone (dust precipitator) removed for repair / Midway area, Minneapolis, Minnesota 1974

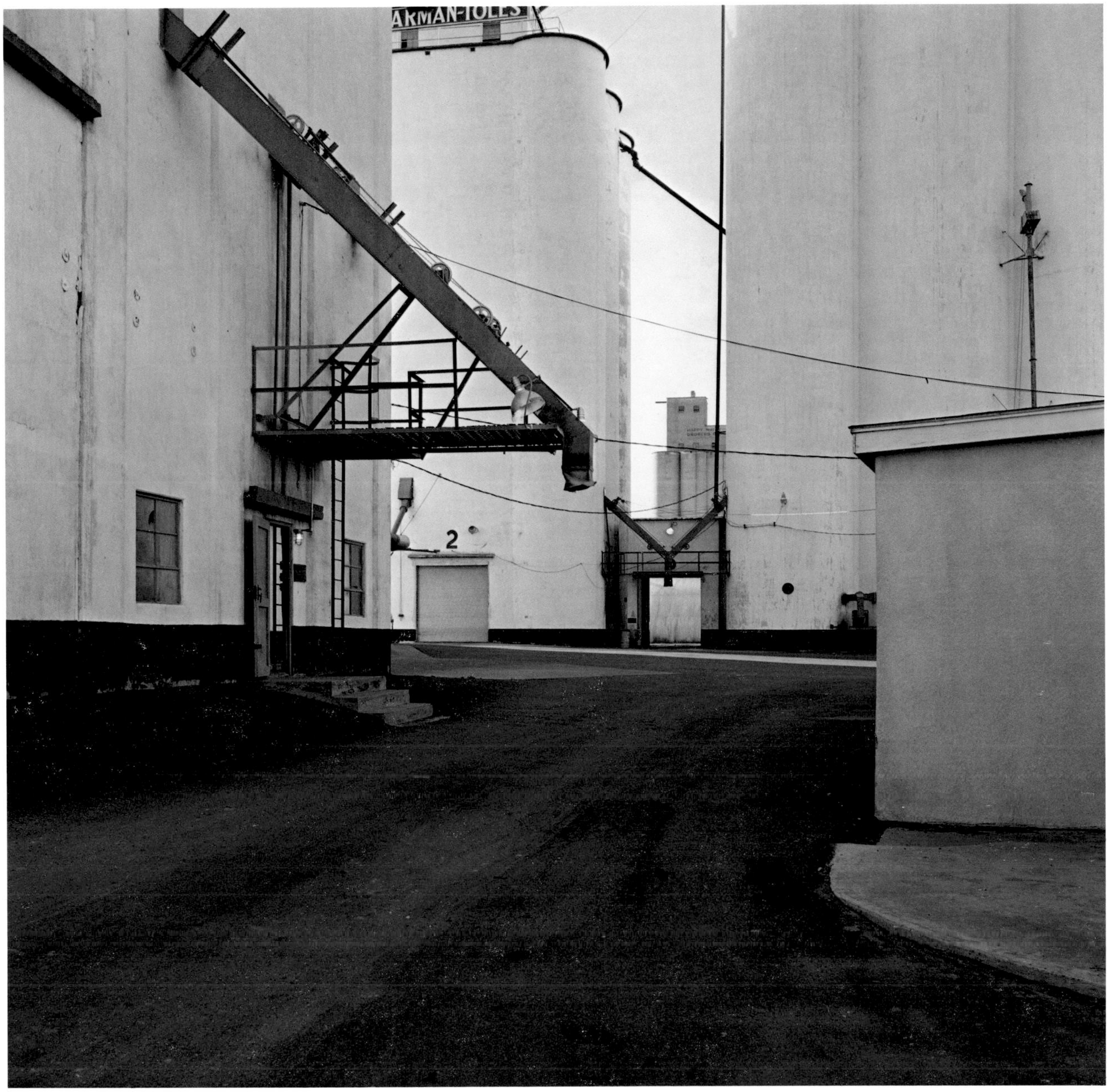

Yard of Harmon-Toles elevator complex / Happy, Texas 1975

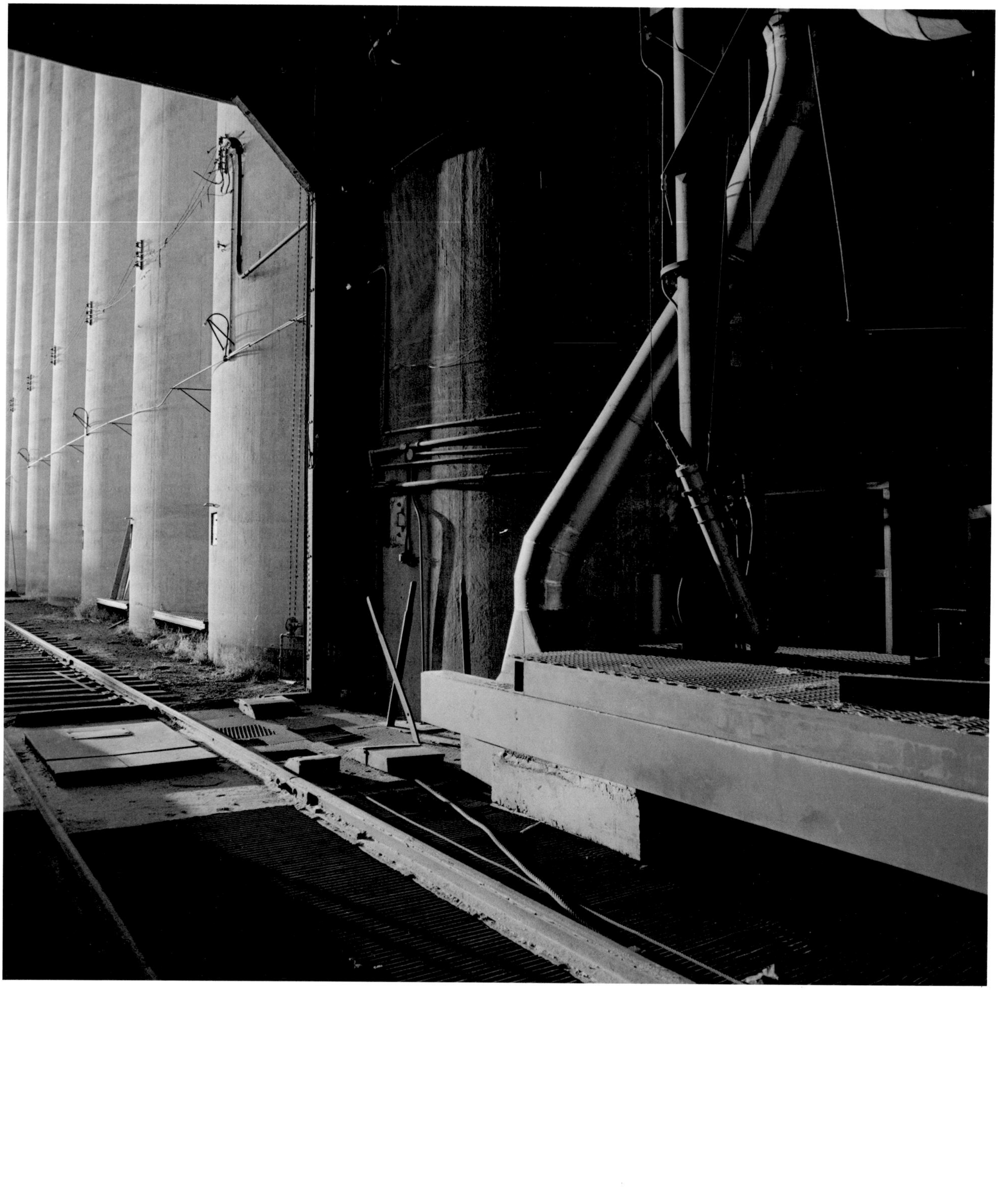

Railcar loading shed / Midway area, Minneapolis, Minnesota 1974

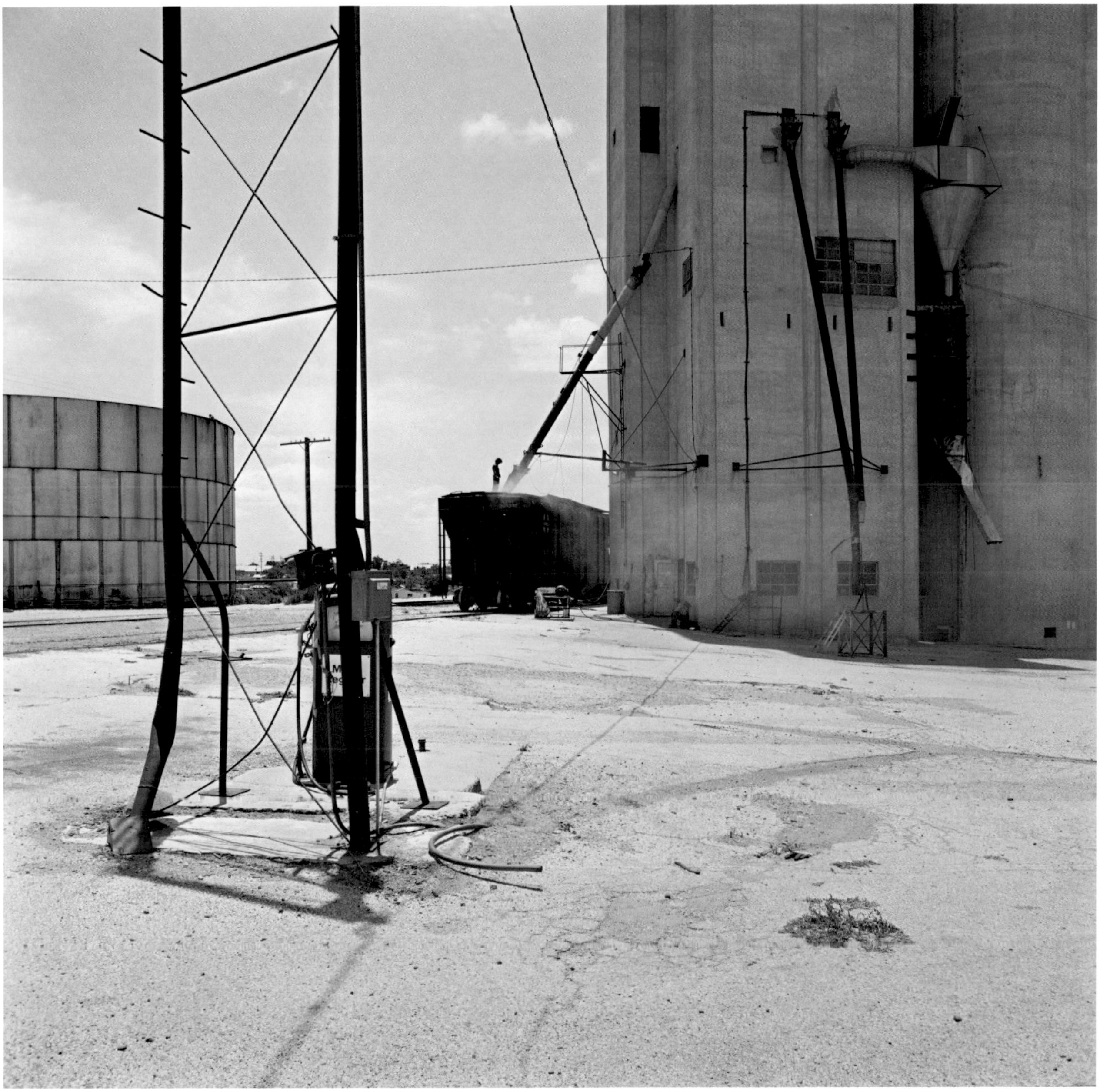

Loading a grain car / Goodpasture elevator, Brownfield, Texas 1975

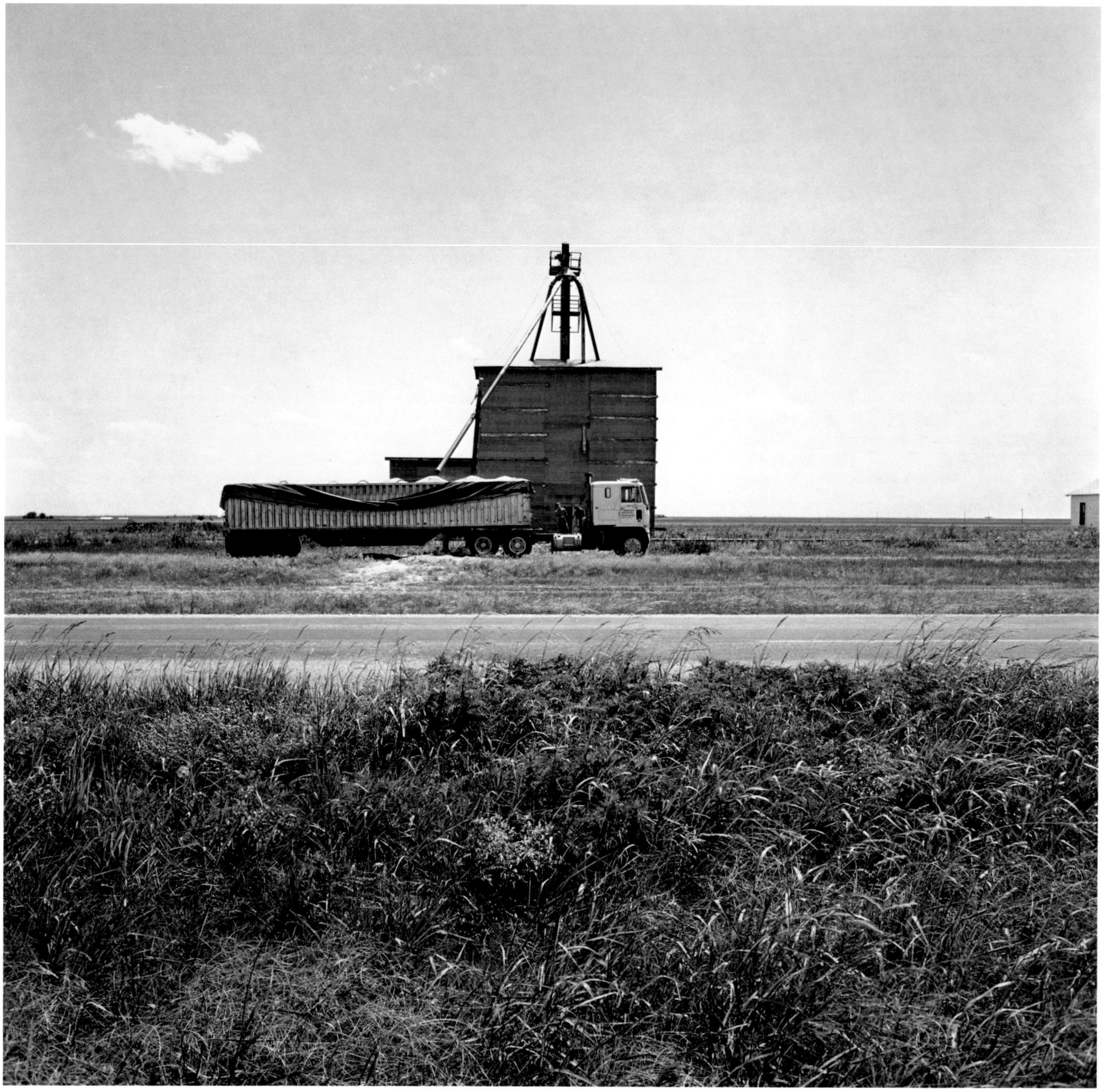

Loading a grain truck / near Roscoe, Texas 1975

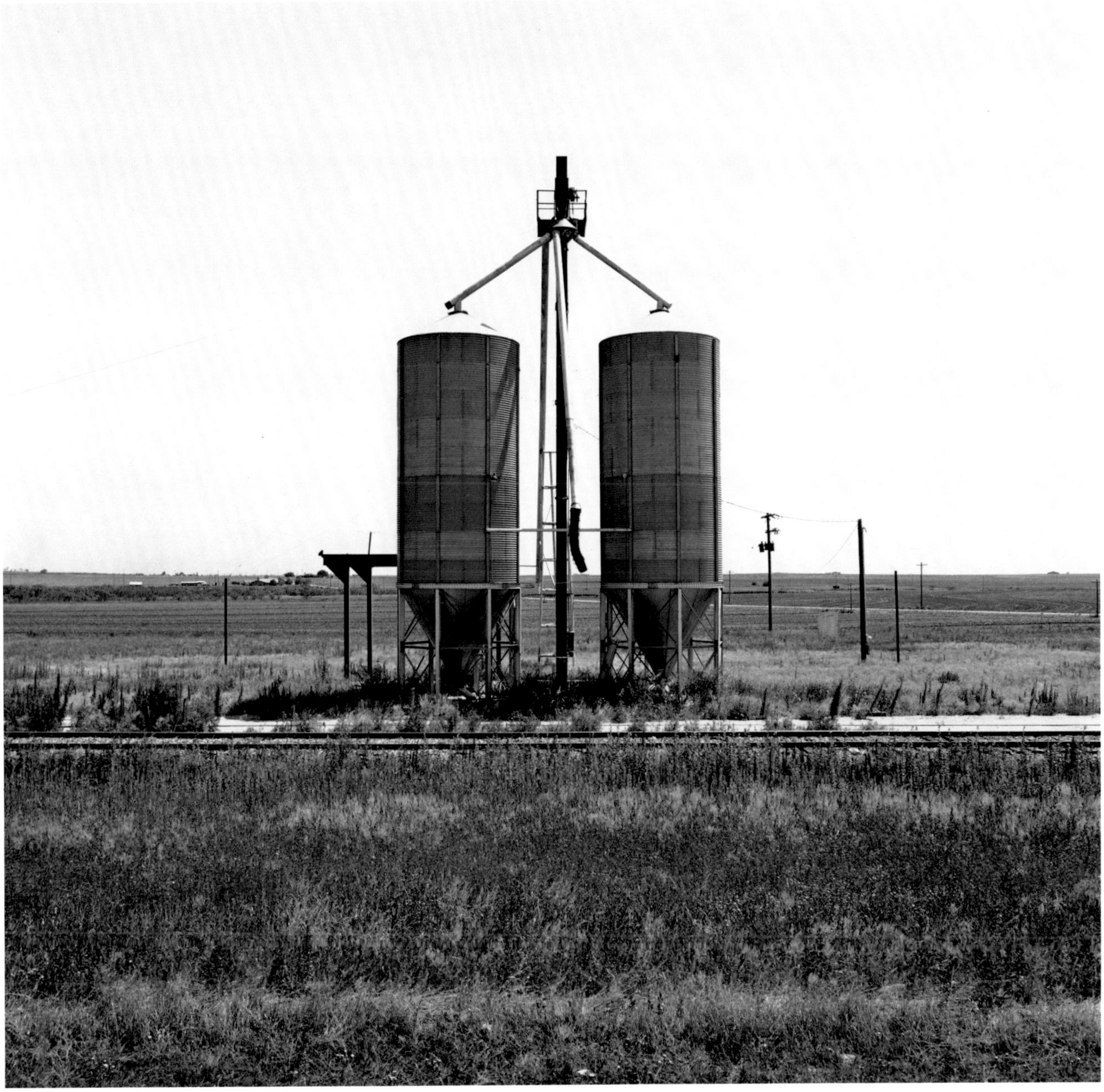

Grain bins and "leg" (vertical bucket conveyor) / near Roscoe, Texas 1975

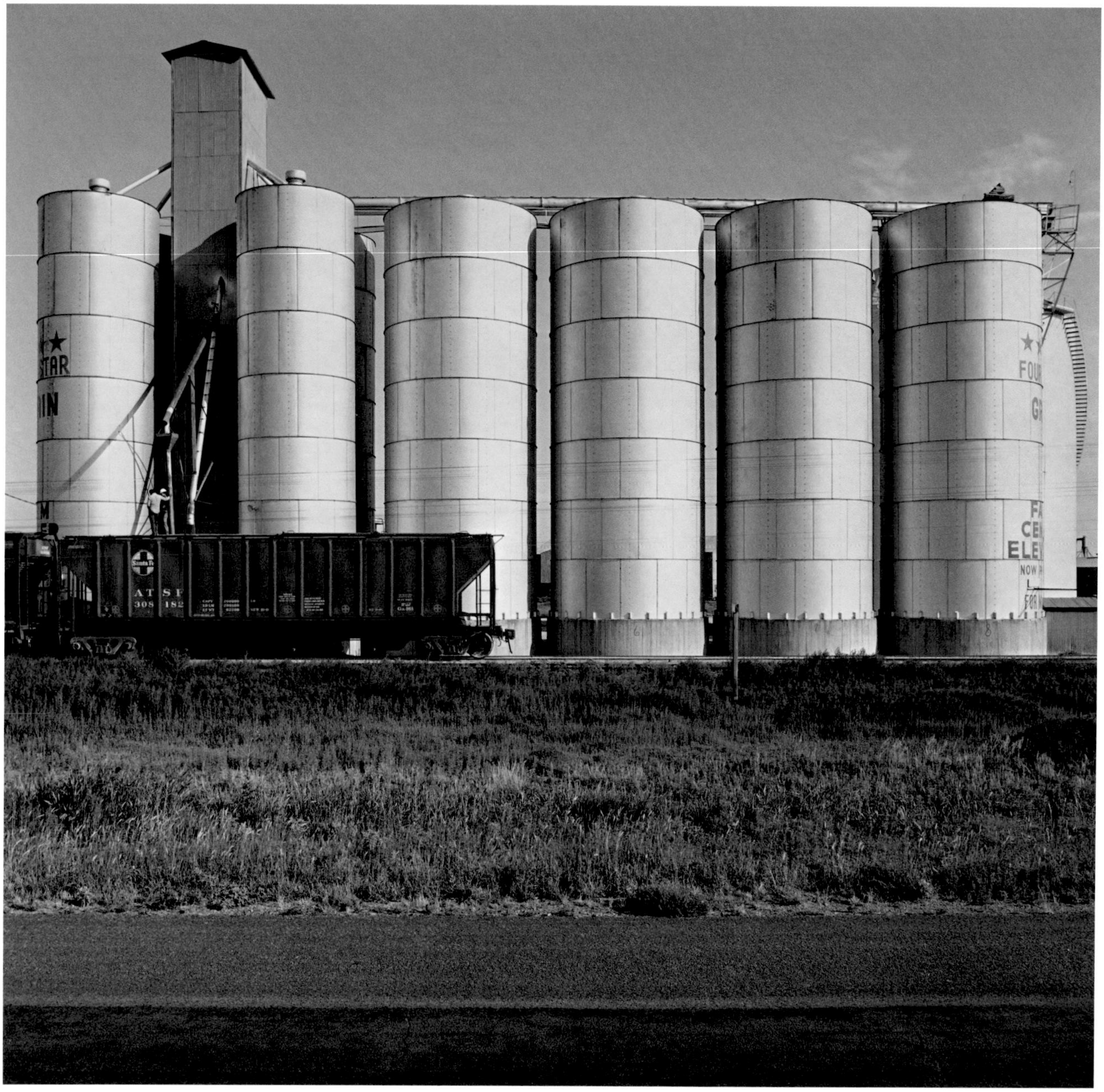

Steel elevator / near Ropesville, Texas 1975

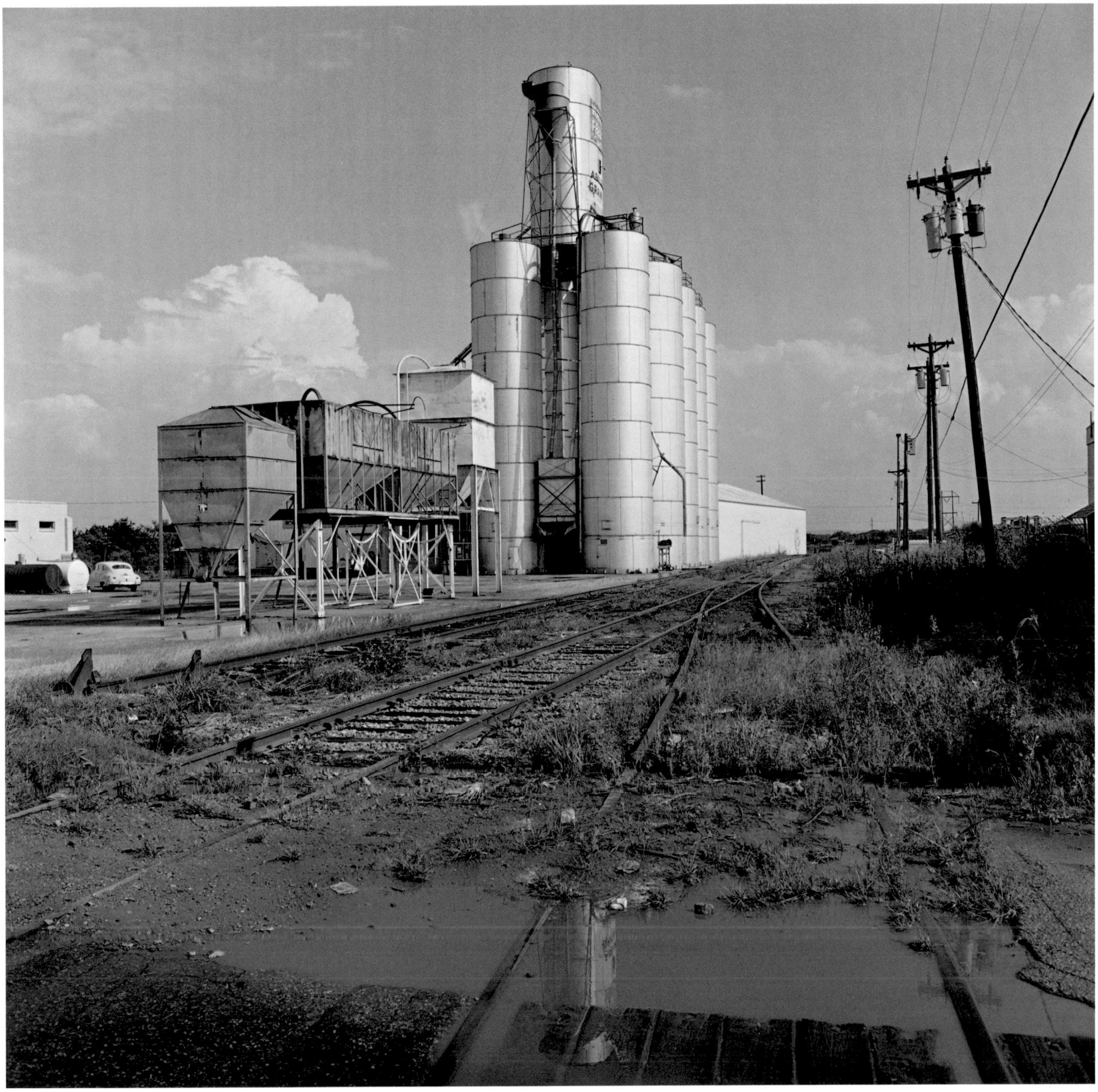

Steel elevator and feed mill / Abilene, Texas 1975

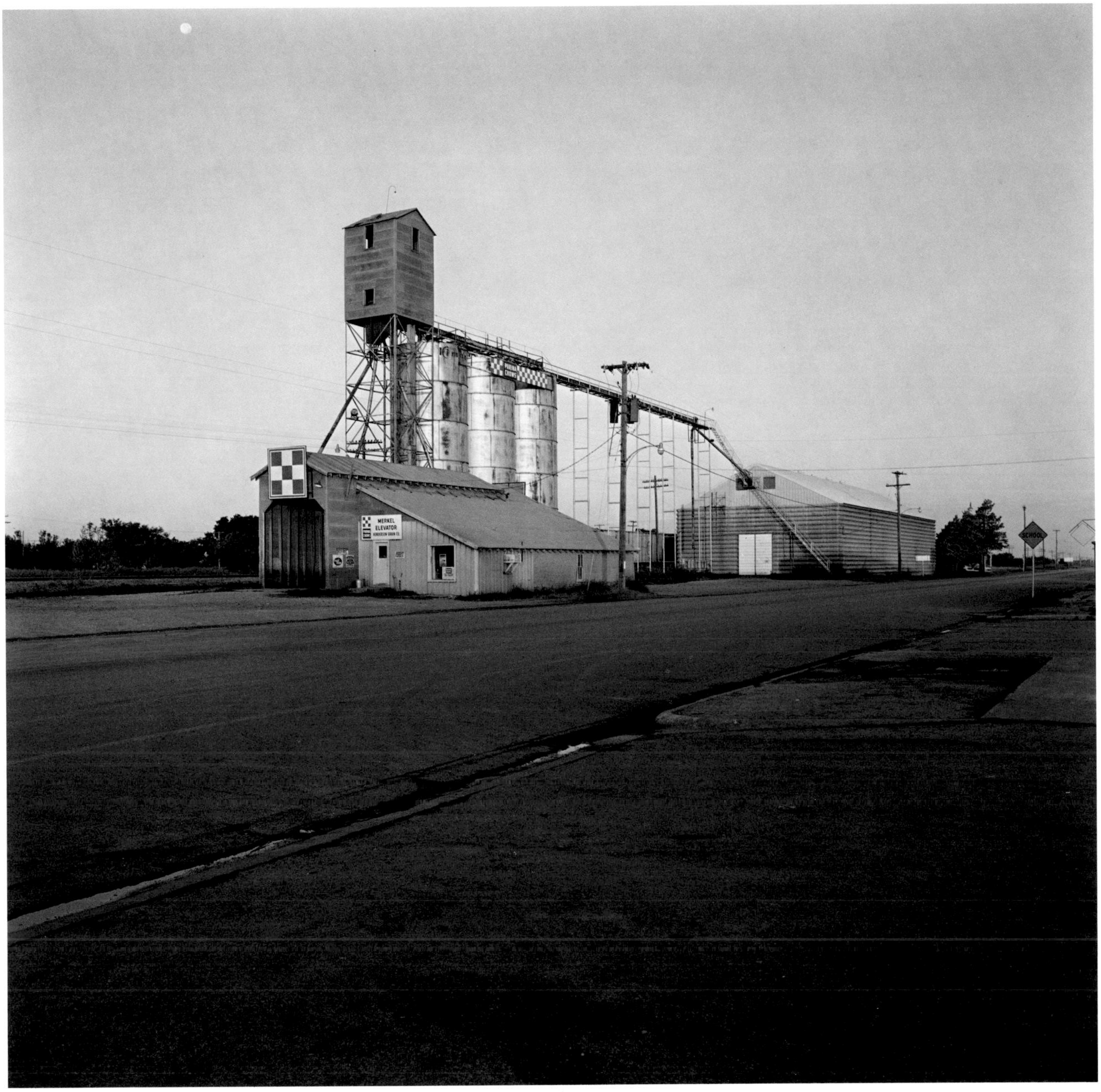

Steel elevator and shed / near Abilene, Texas 1975

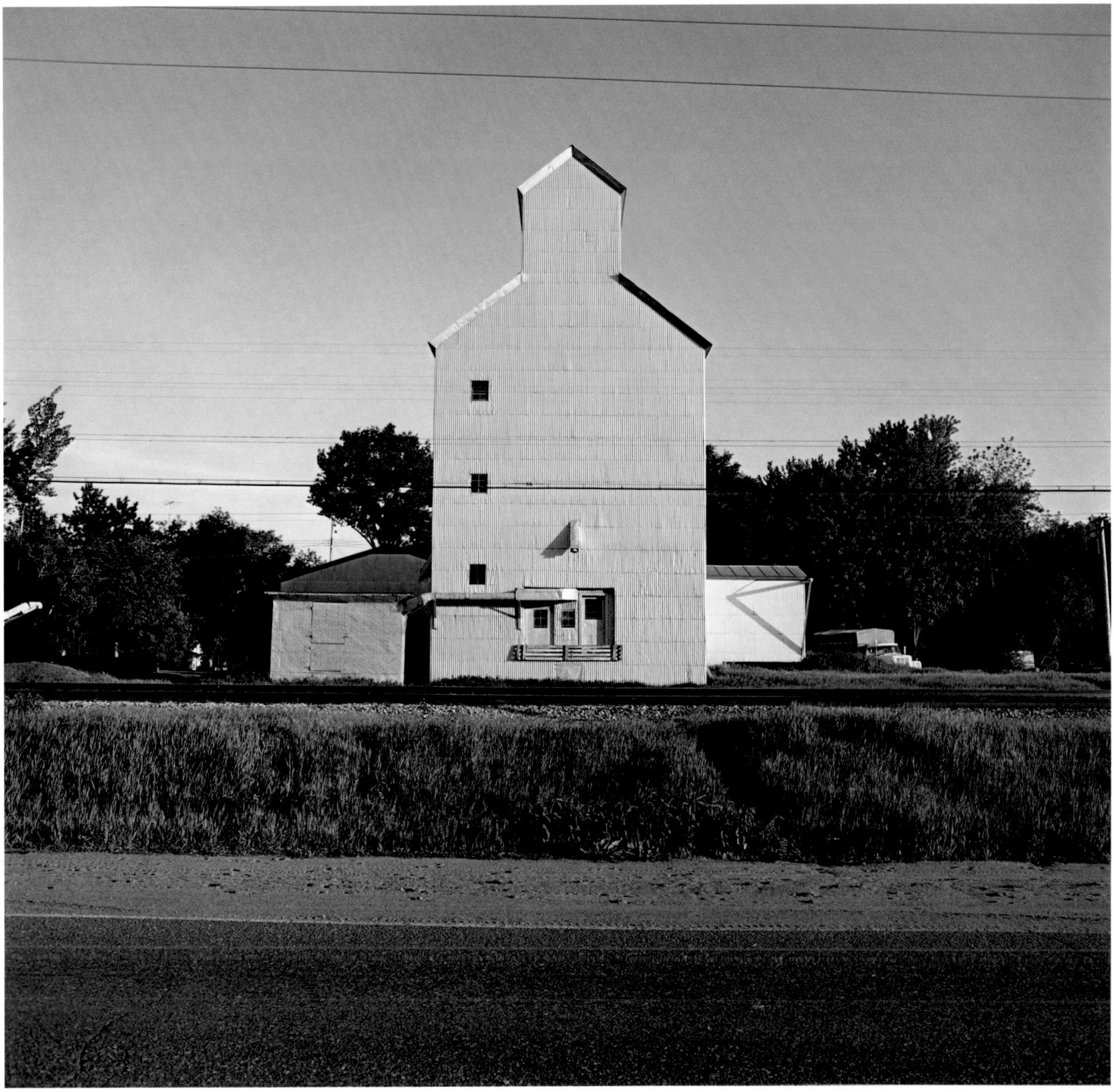

Grain elevator / Bay City, Wisconsin 1973

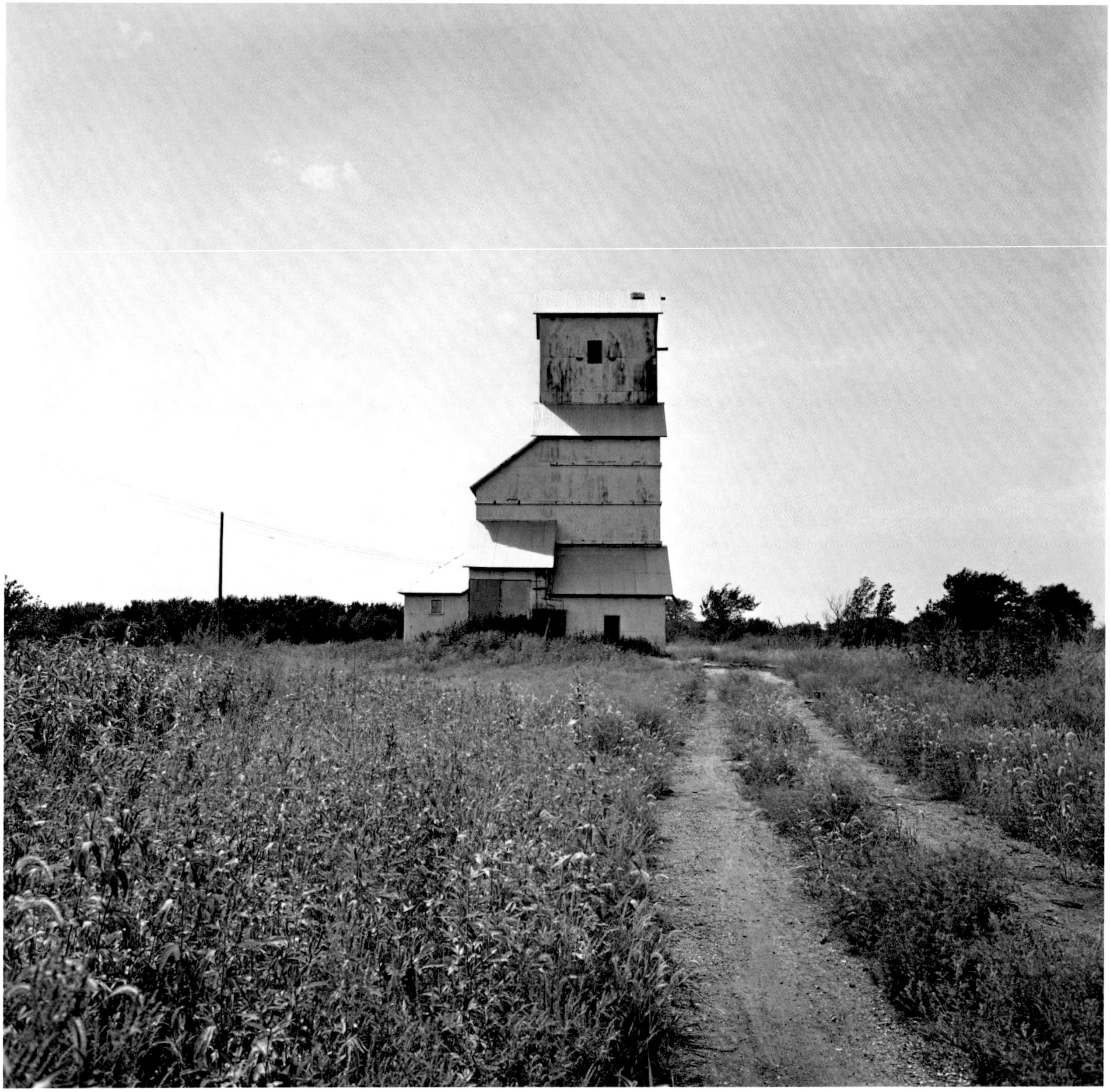

Abandoned grain elevator / Homewood, Kansas 1973

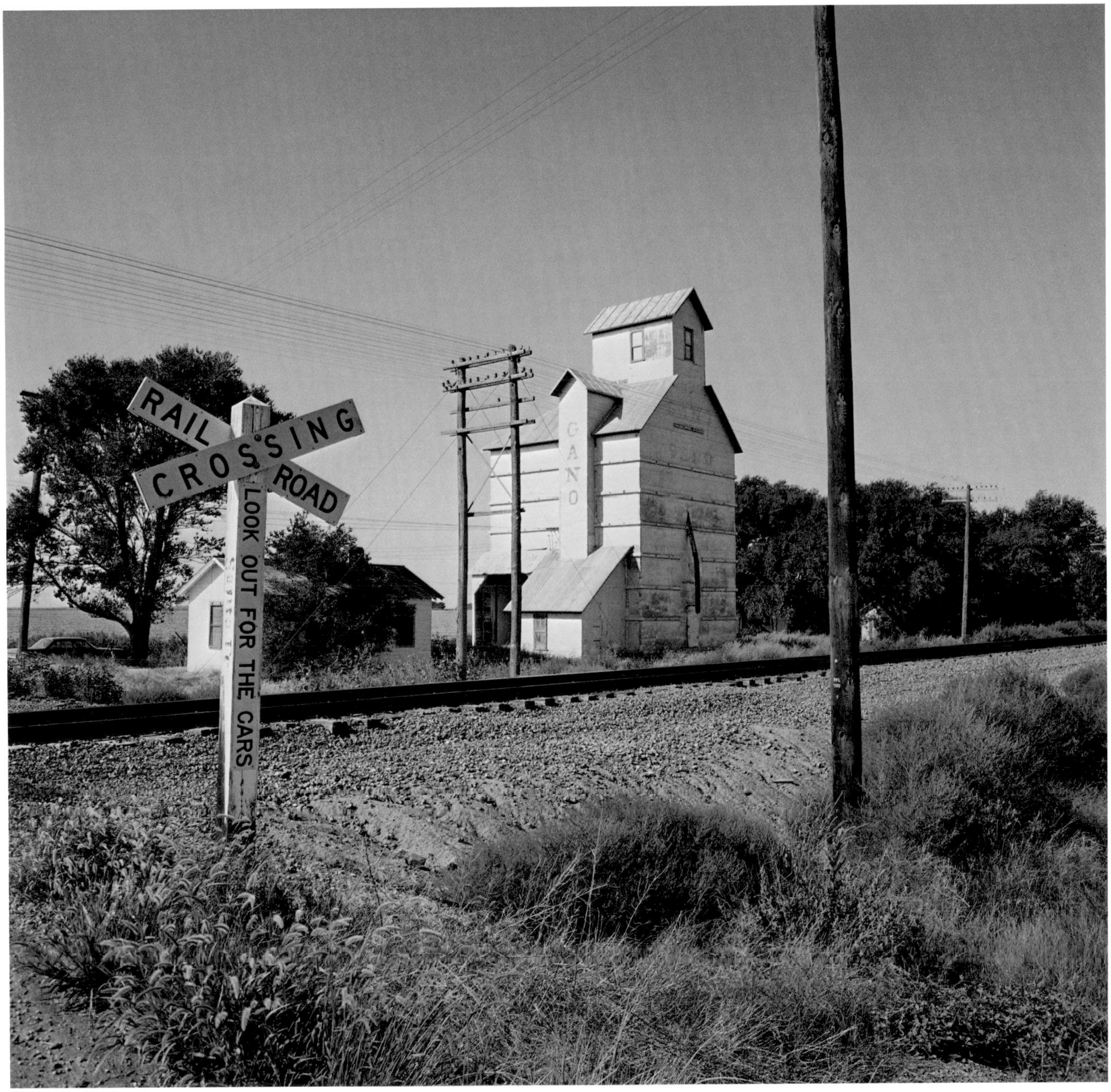

Grain elevator / near Kinsley, Kansas 1973

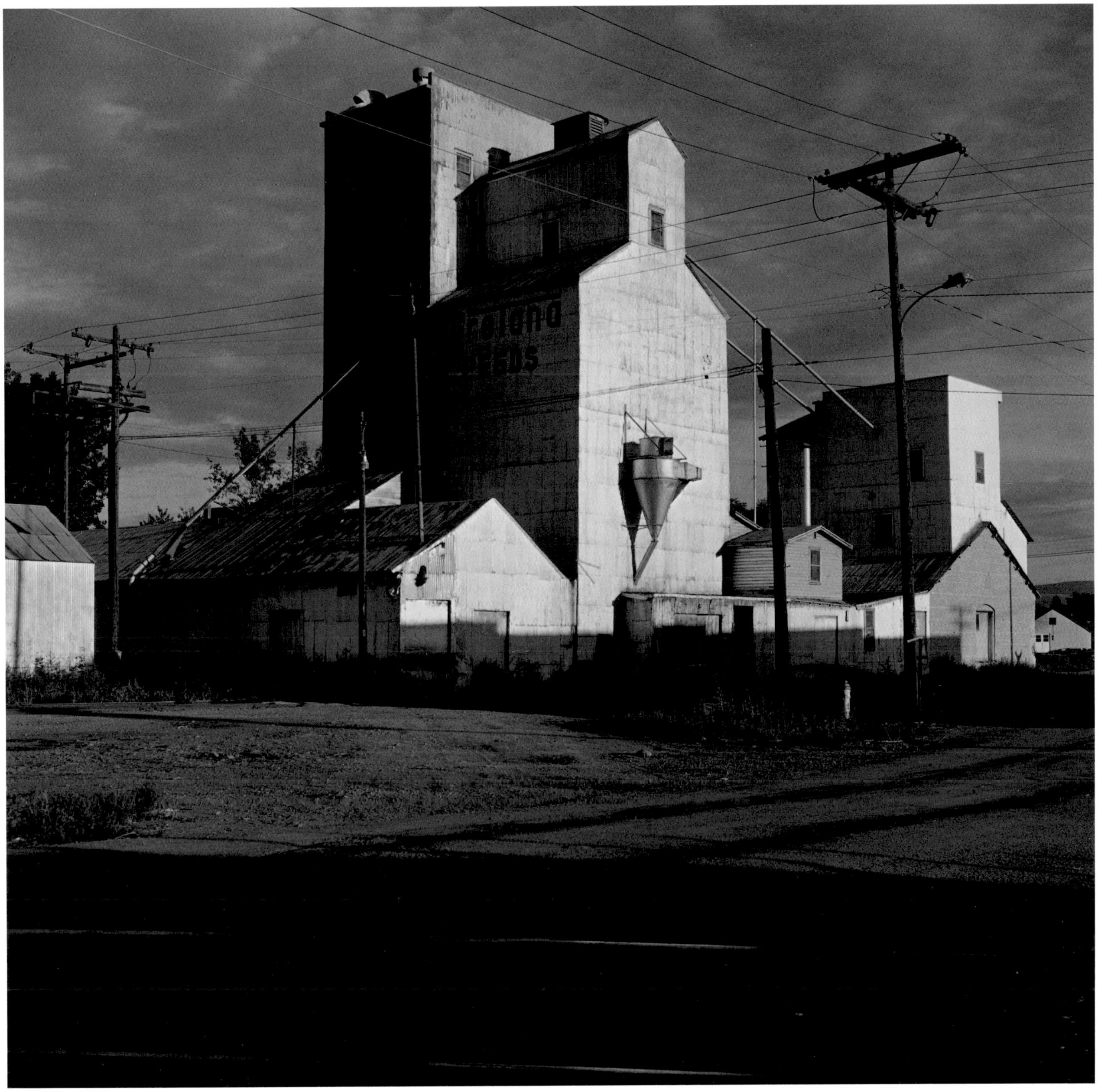

Grain elevator and feed mill / Missoula, Montana 1974

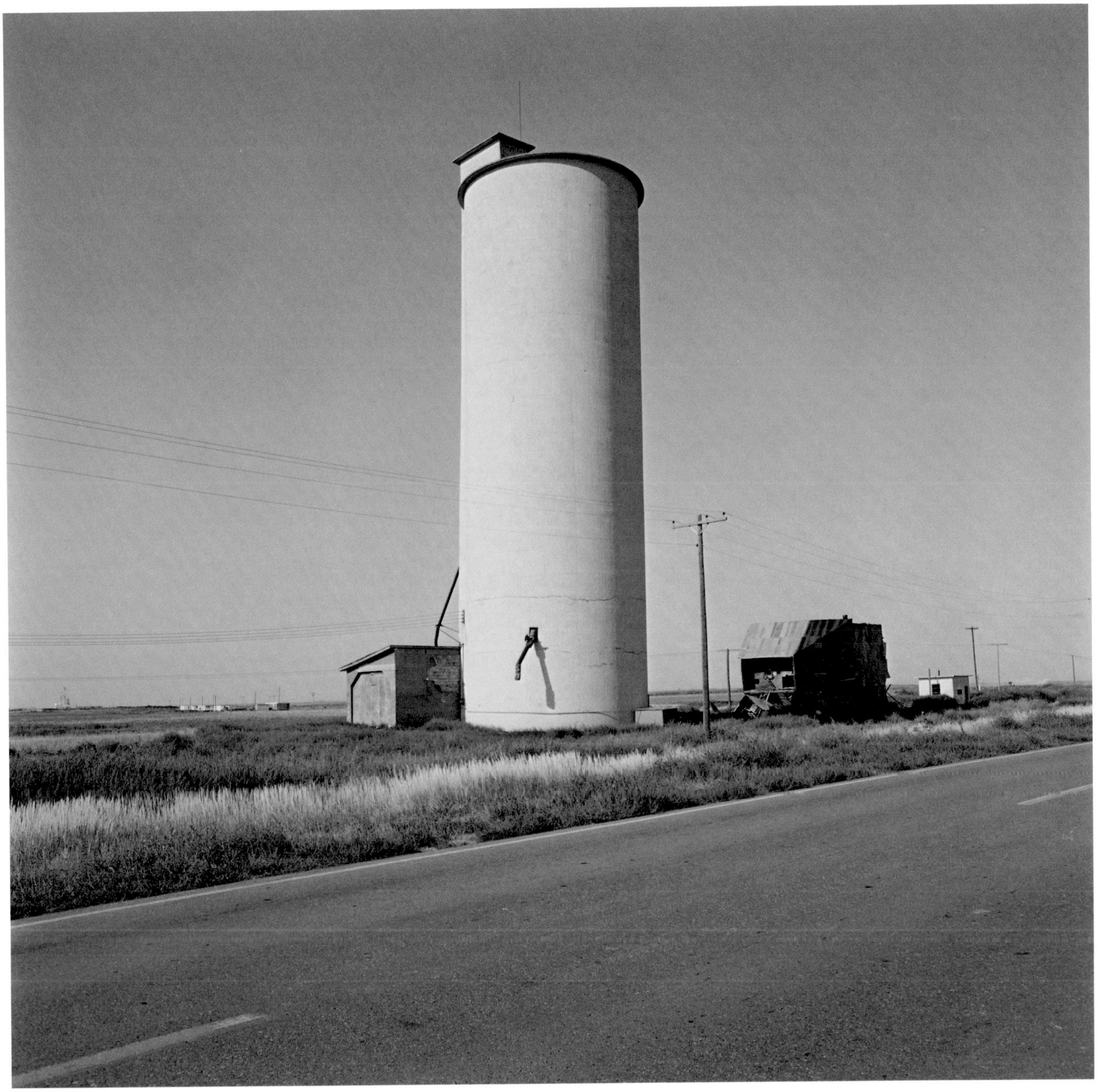

Abandoned grain elevator / near Copeland, Kansas 1973

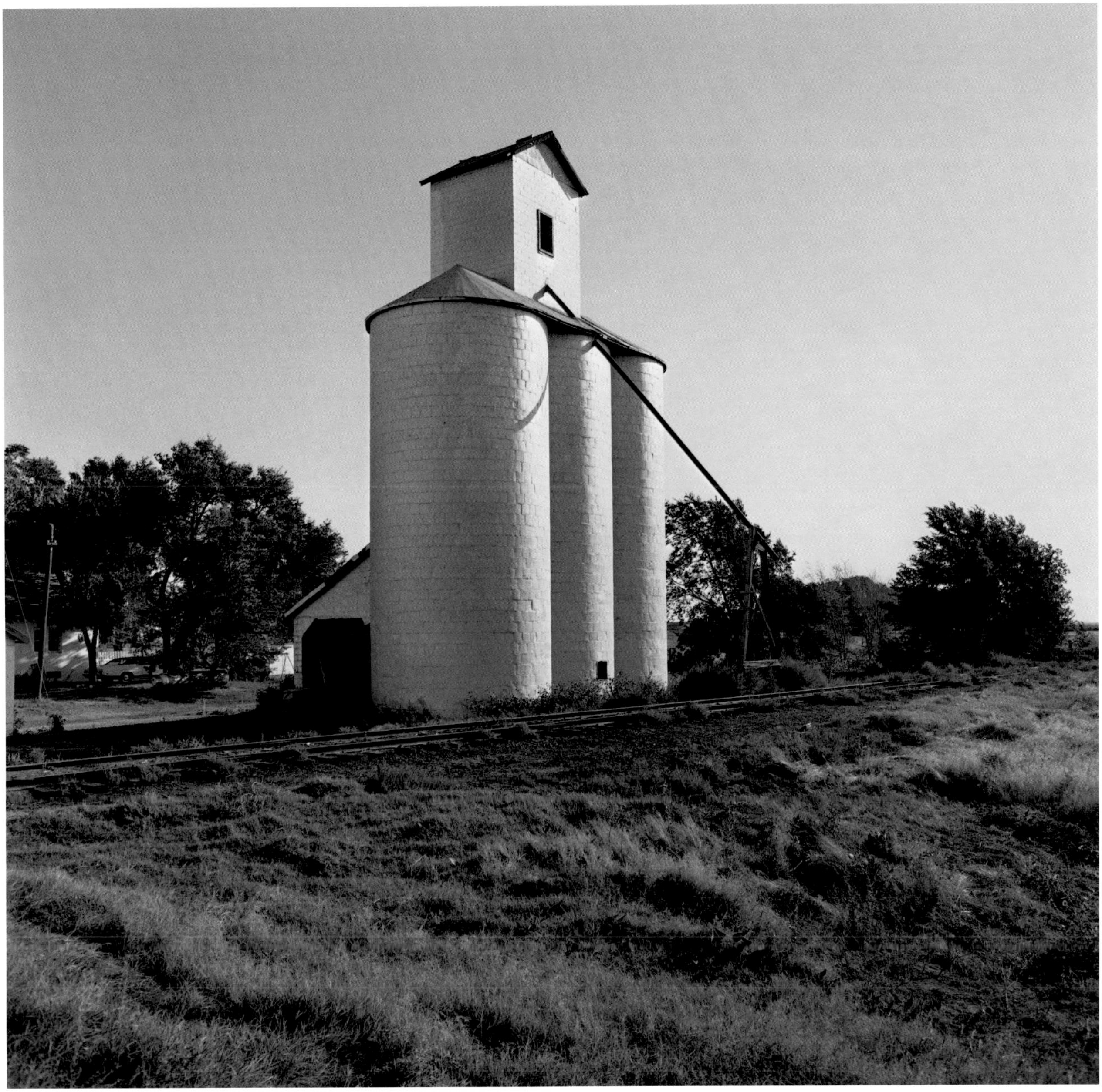

Grain elevator / Bellefont, Kansas 1973

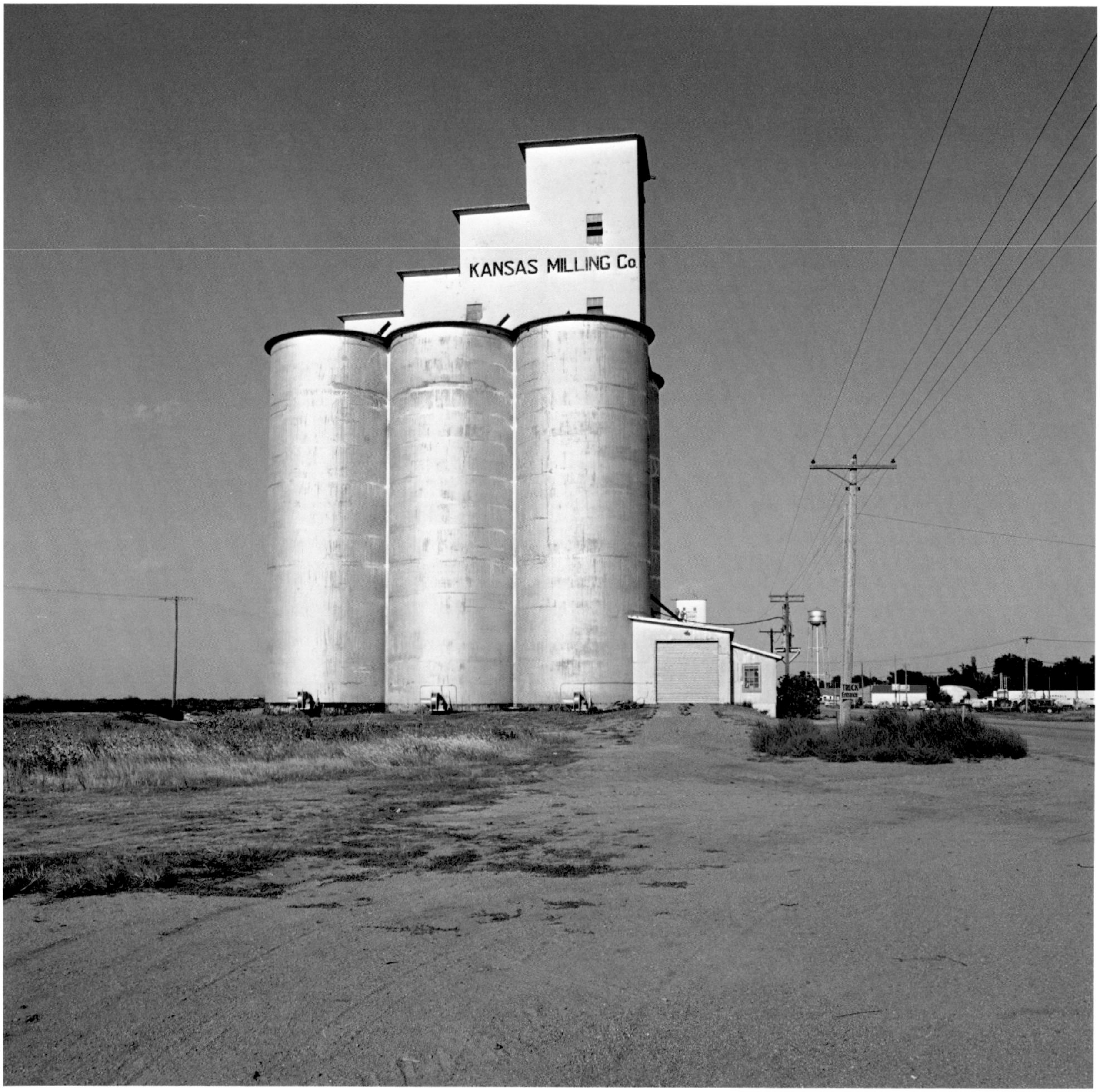

Grain elevator / Montezuma, Kansas 1973

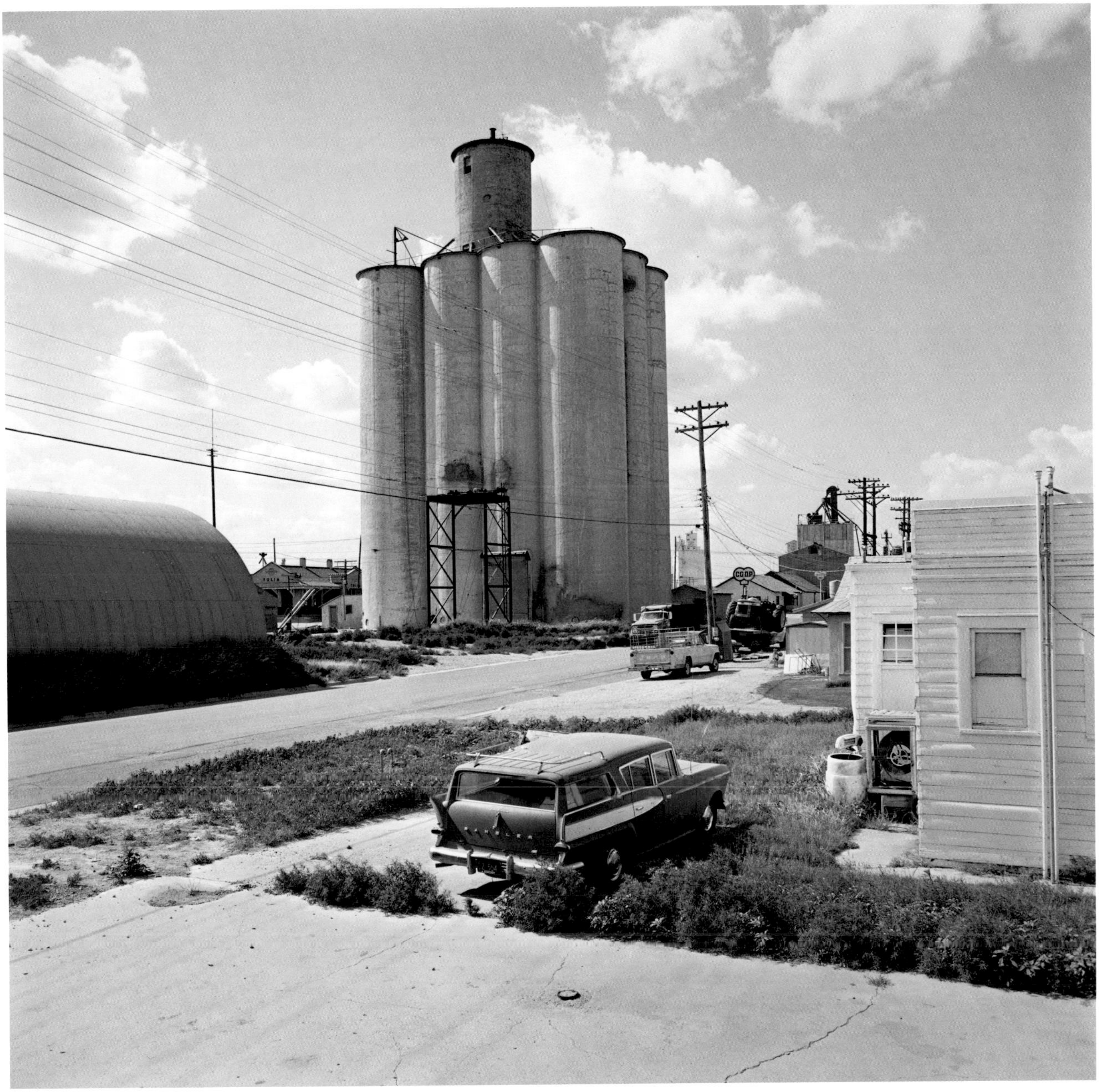

Grain elevator / Tulia, Texas 1975

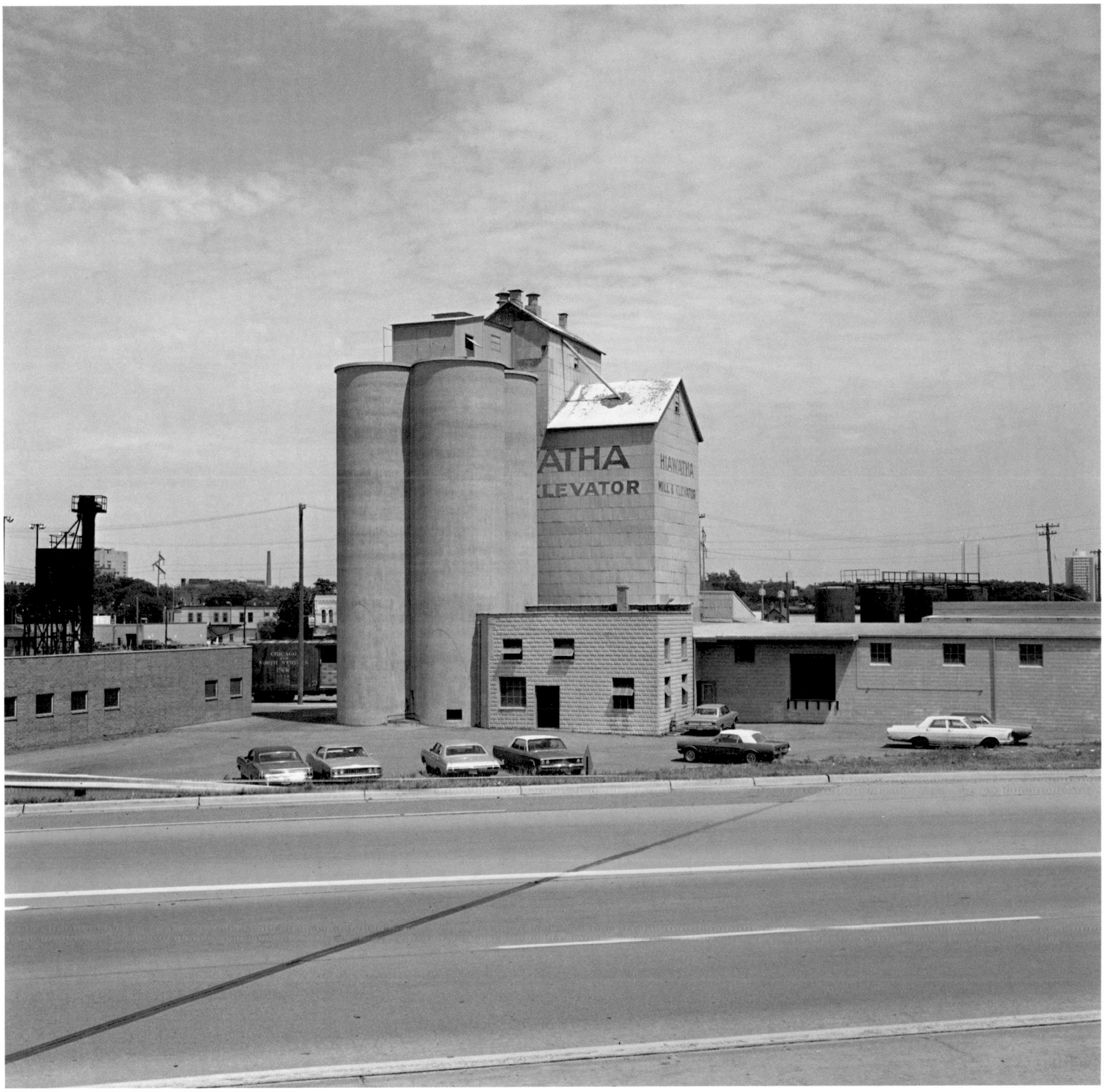

Hiawatha Elevator / Minneapolis, Minnesota 1974 (Destroyed in explosion in 1979)

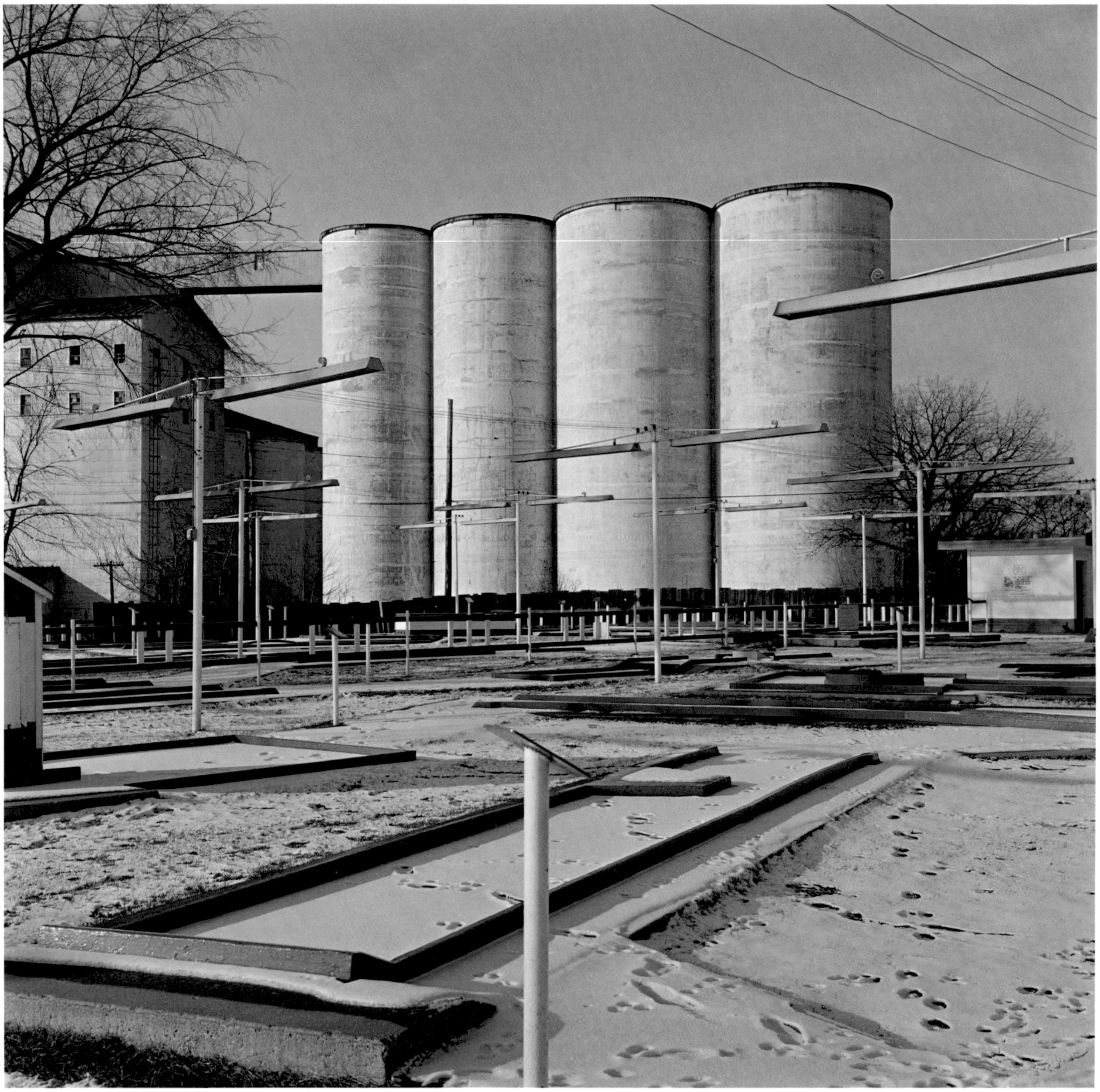

Grain elevator / Minneapolis, Minnesota 1974

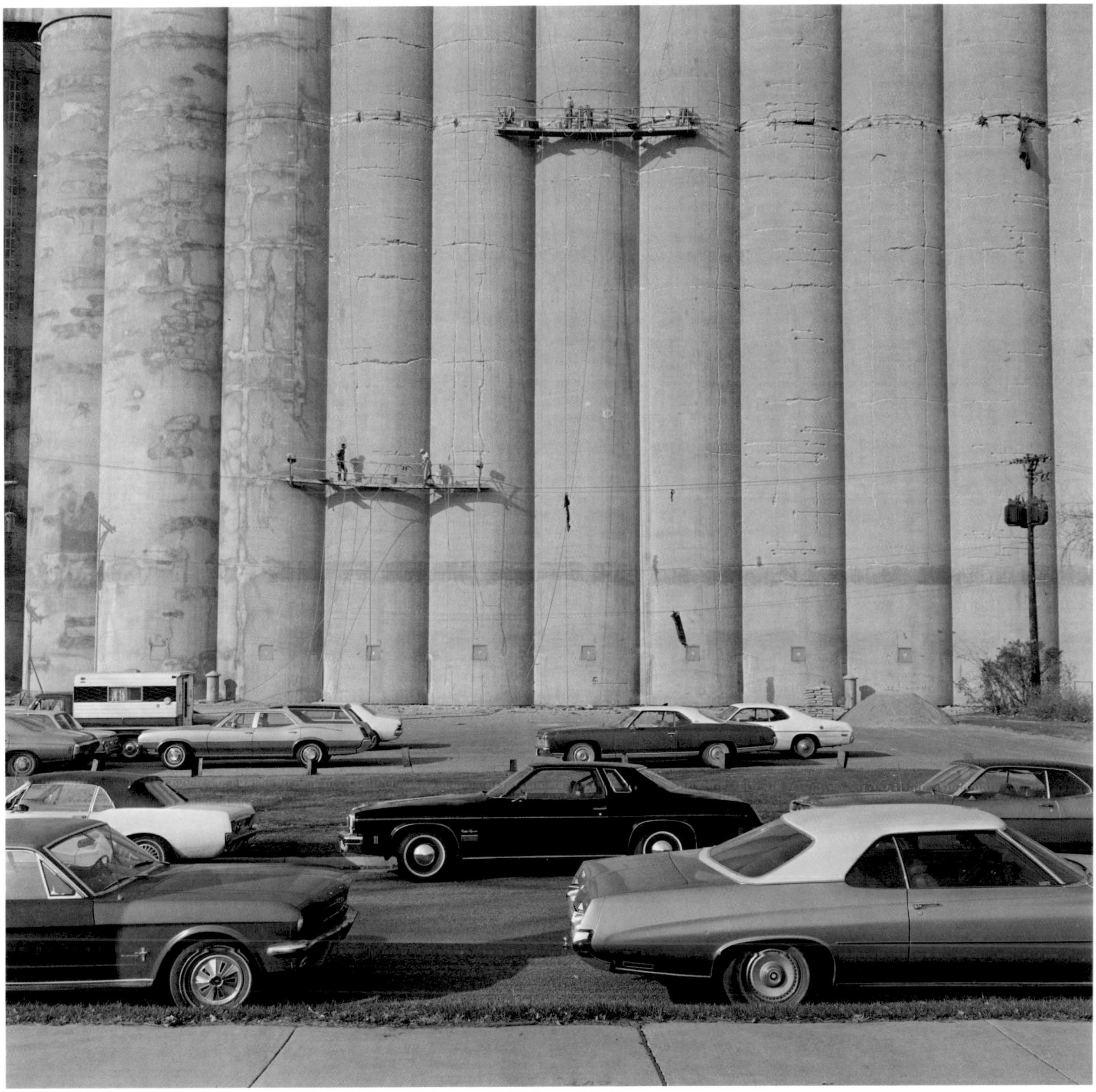

Grain elevator being repaired / Minneapolis, Minnesota 1974

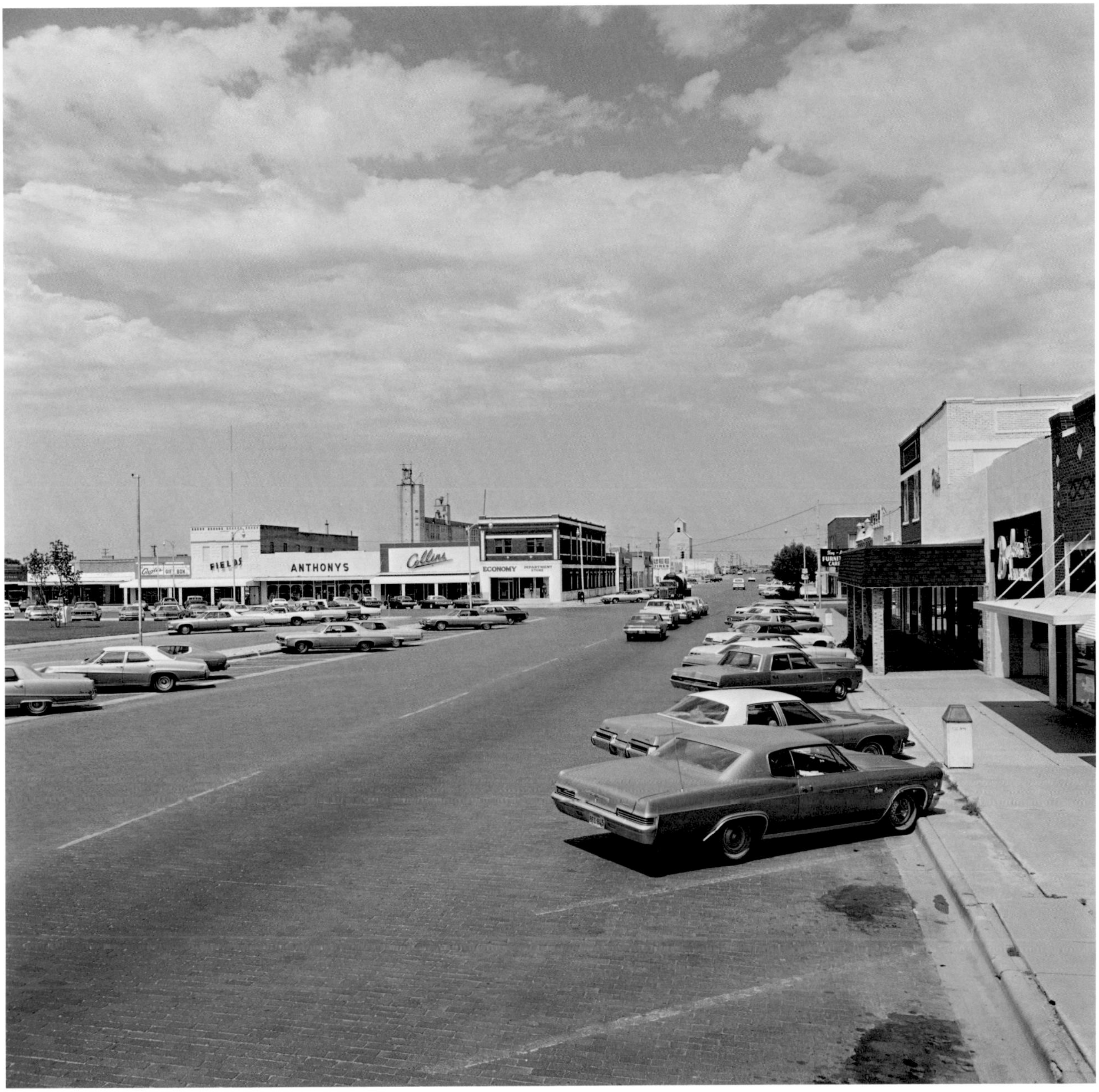

Town view / Brownfield, Texas 1975

Grain elevator and plowed field / Wellington, Kansas 1973

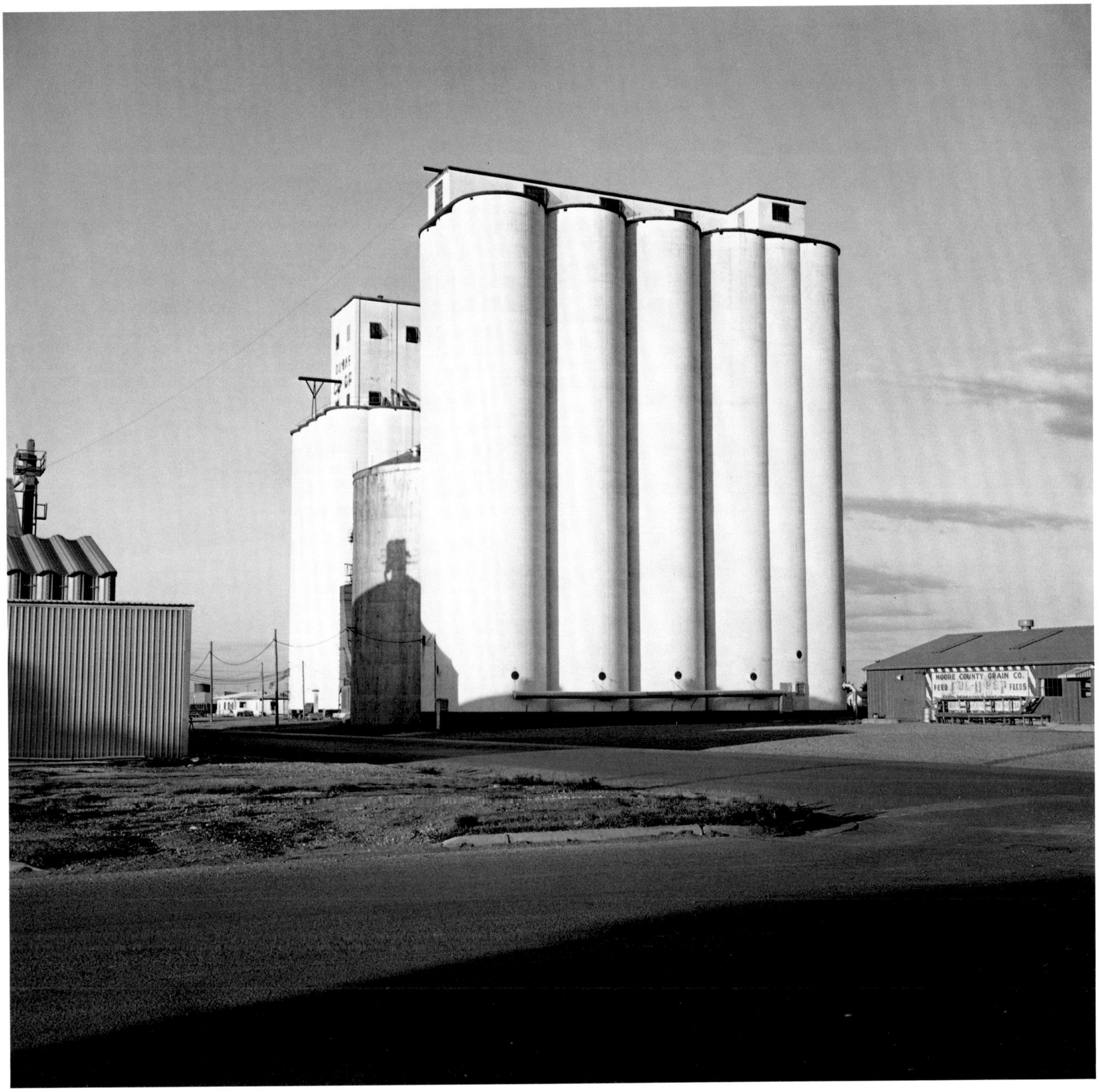

Grain elevator / Dumas, Texas 1973

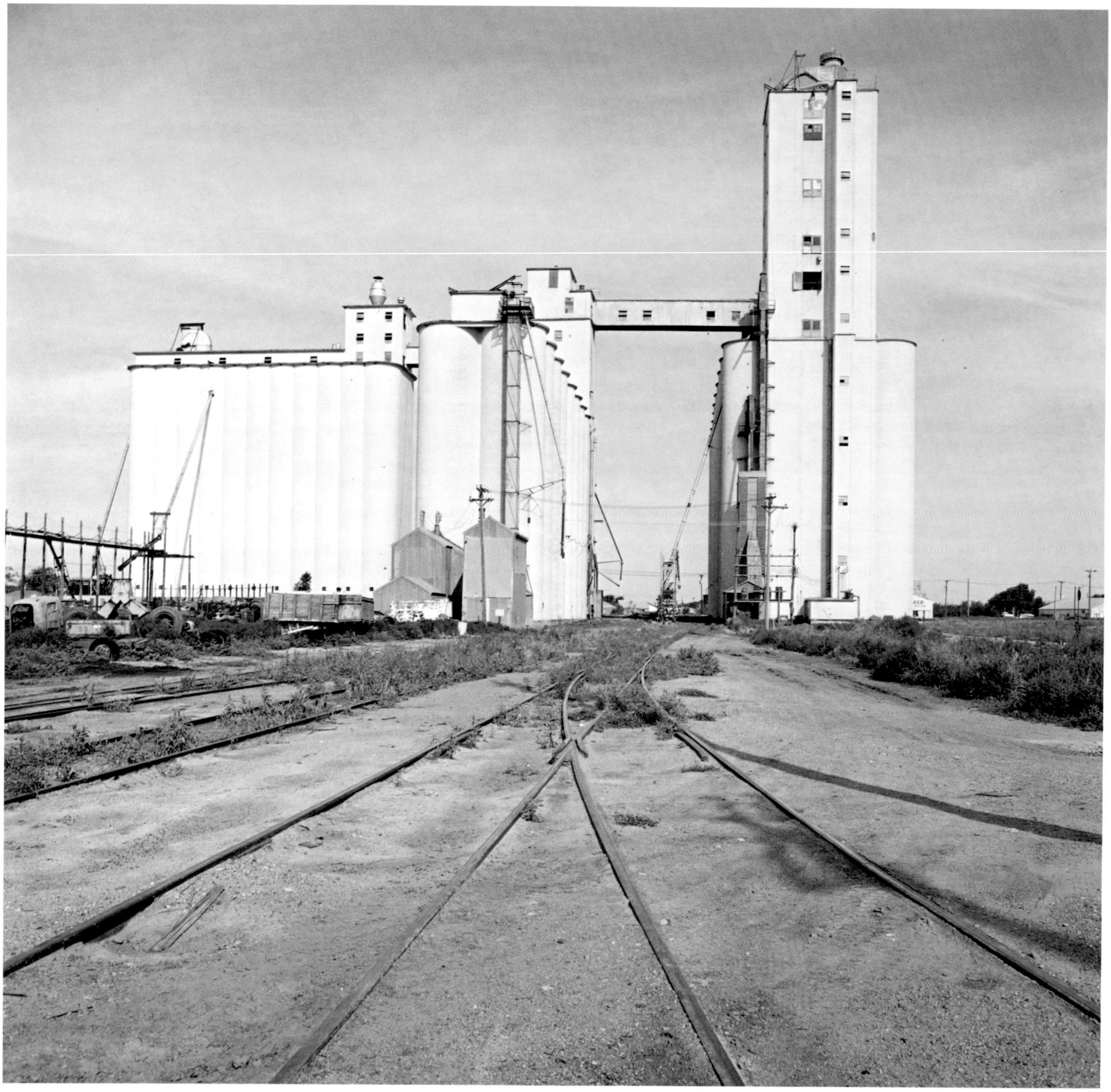

Harvest Queen Elevator / Plainview, Texas 1975

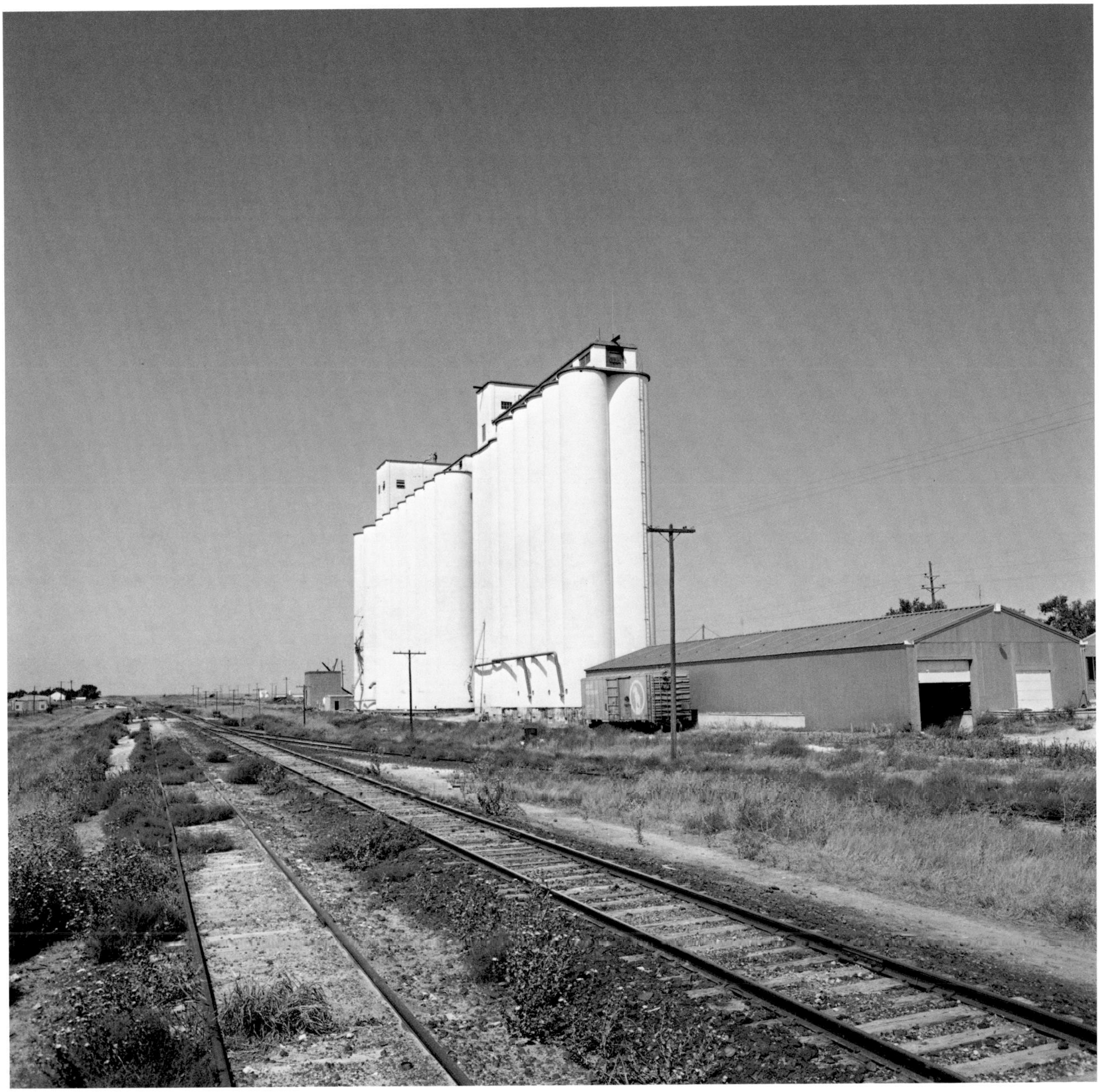

Grain elevators / Follett, Texas 1973

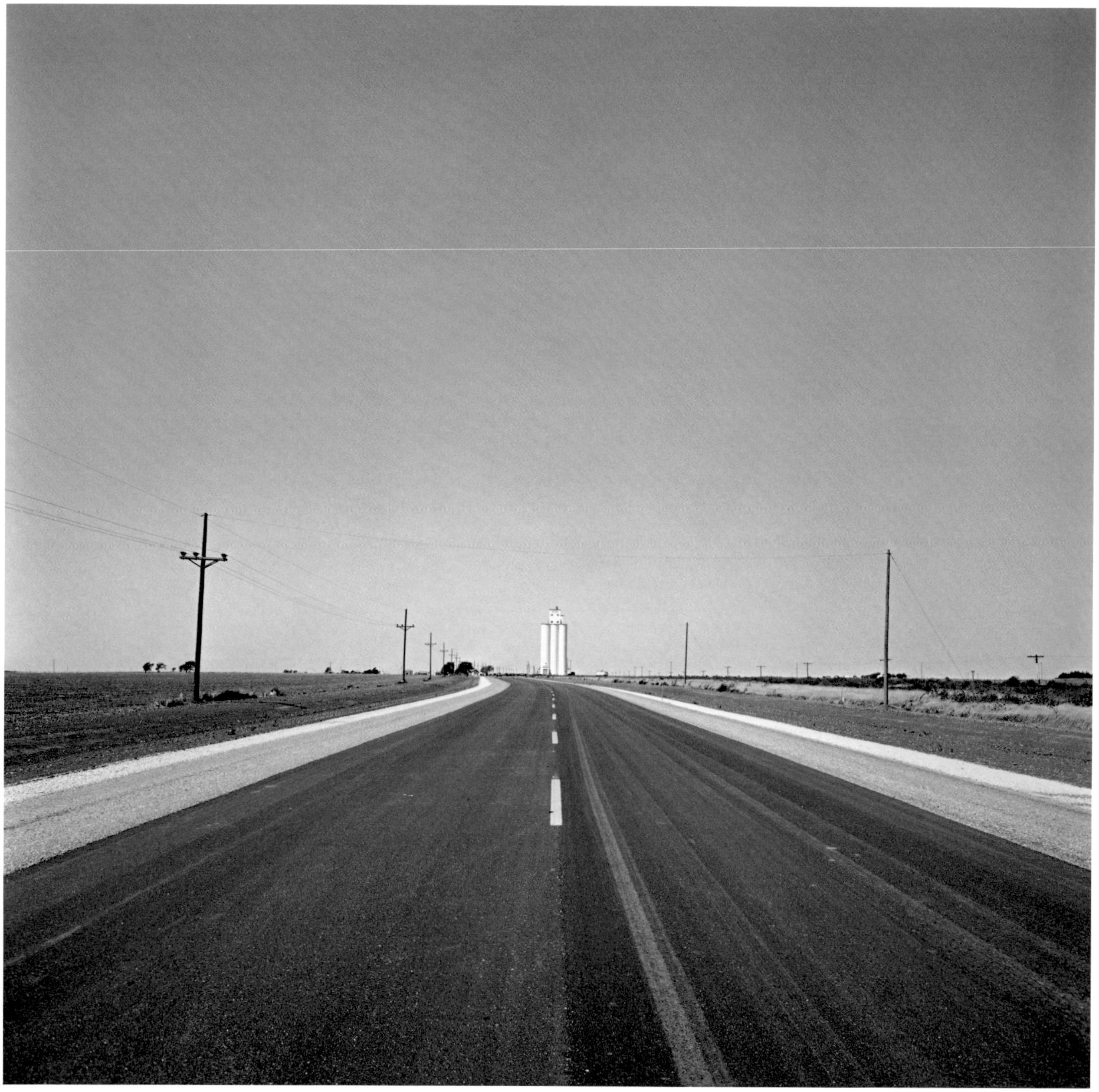

Grain elevator in the bend of the road / Perryton, Texas 1975

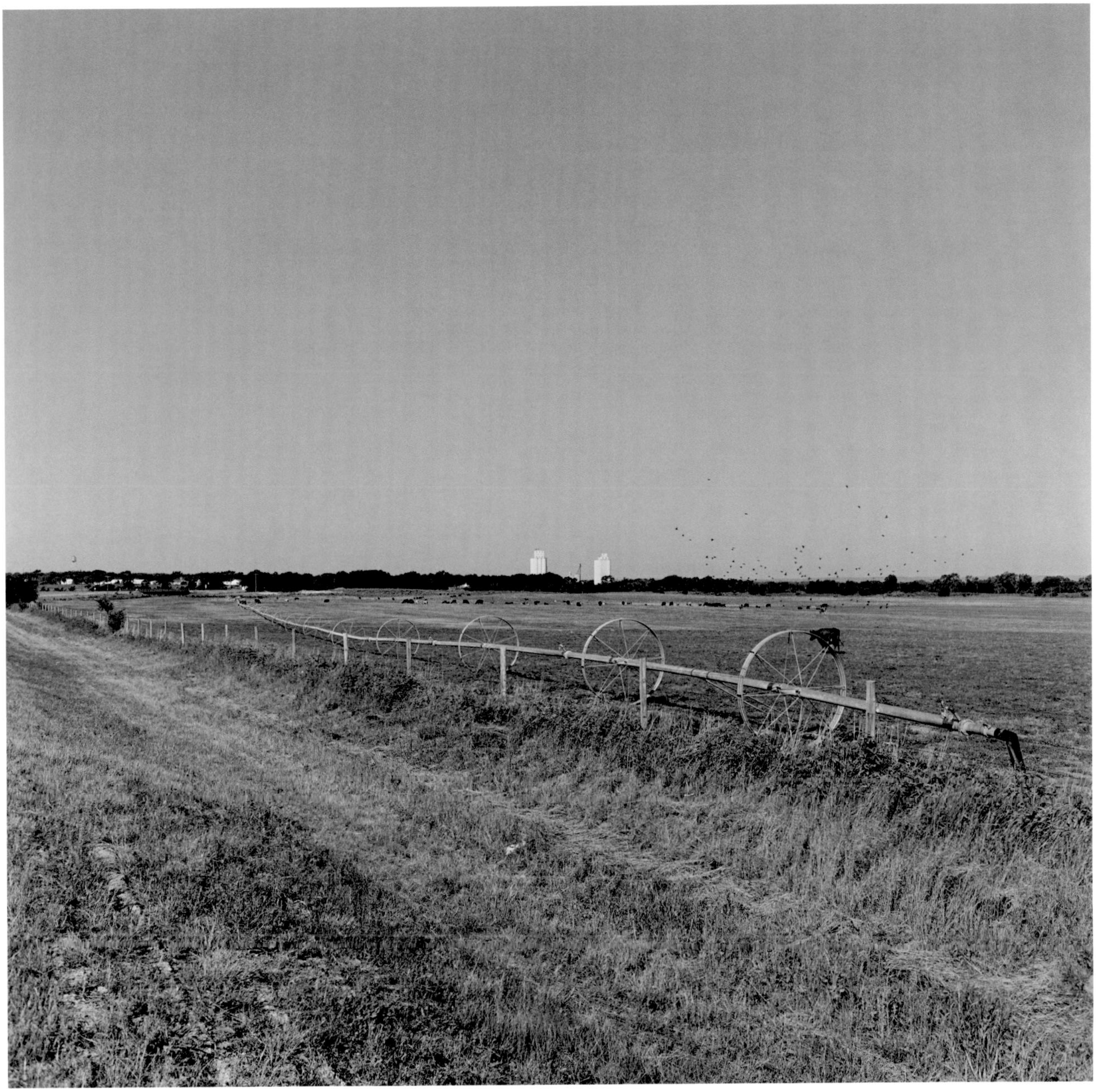

Town view / Woodward, Oklahoma 1973

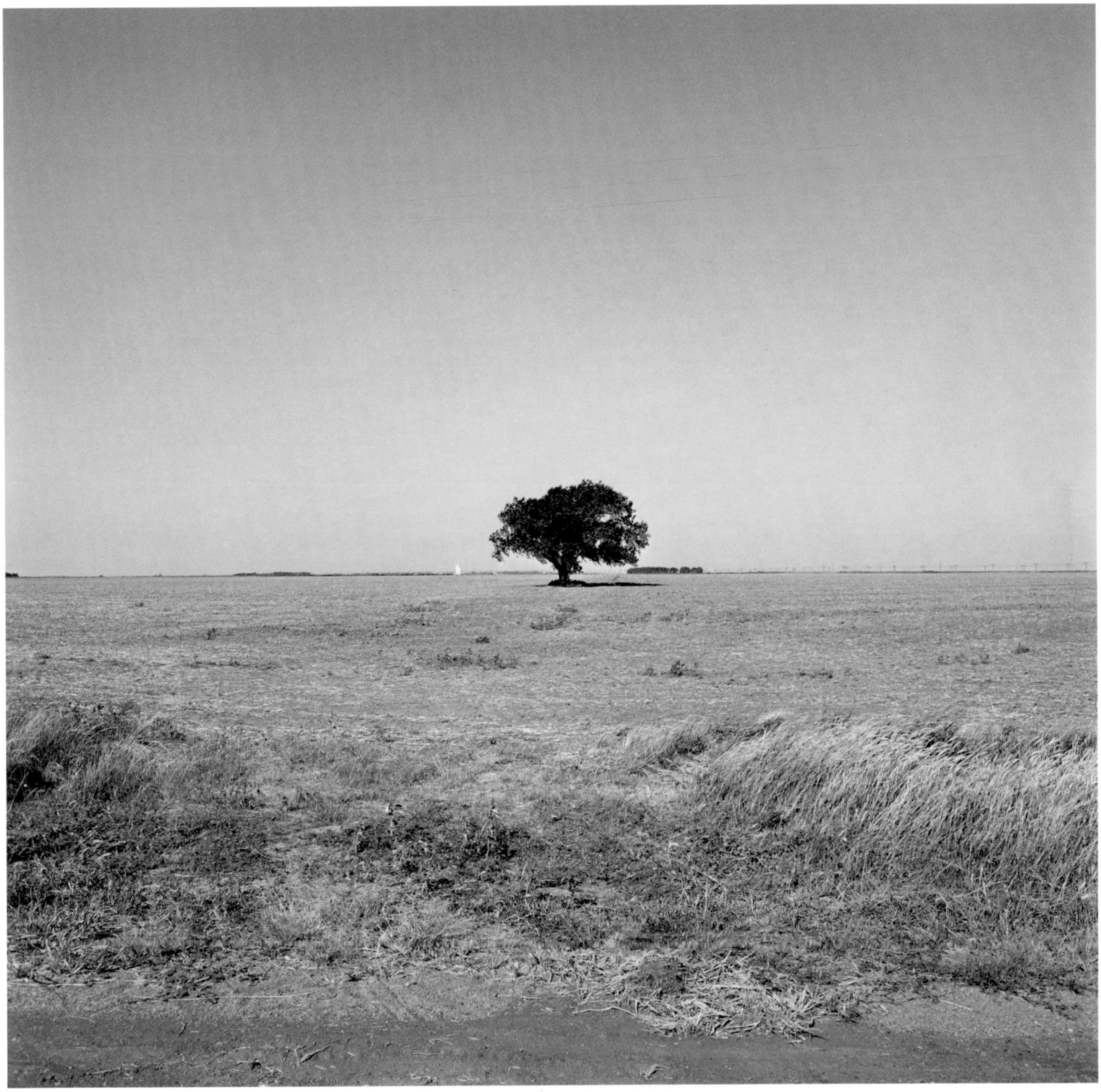

Landscape / near Kinsley, Kansas 1973

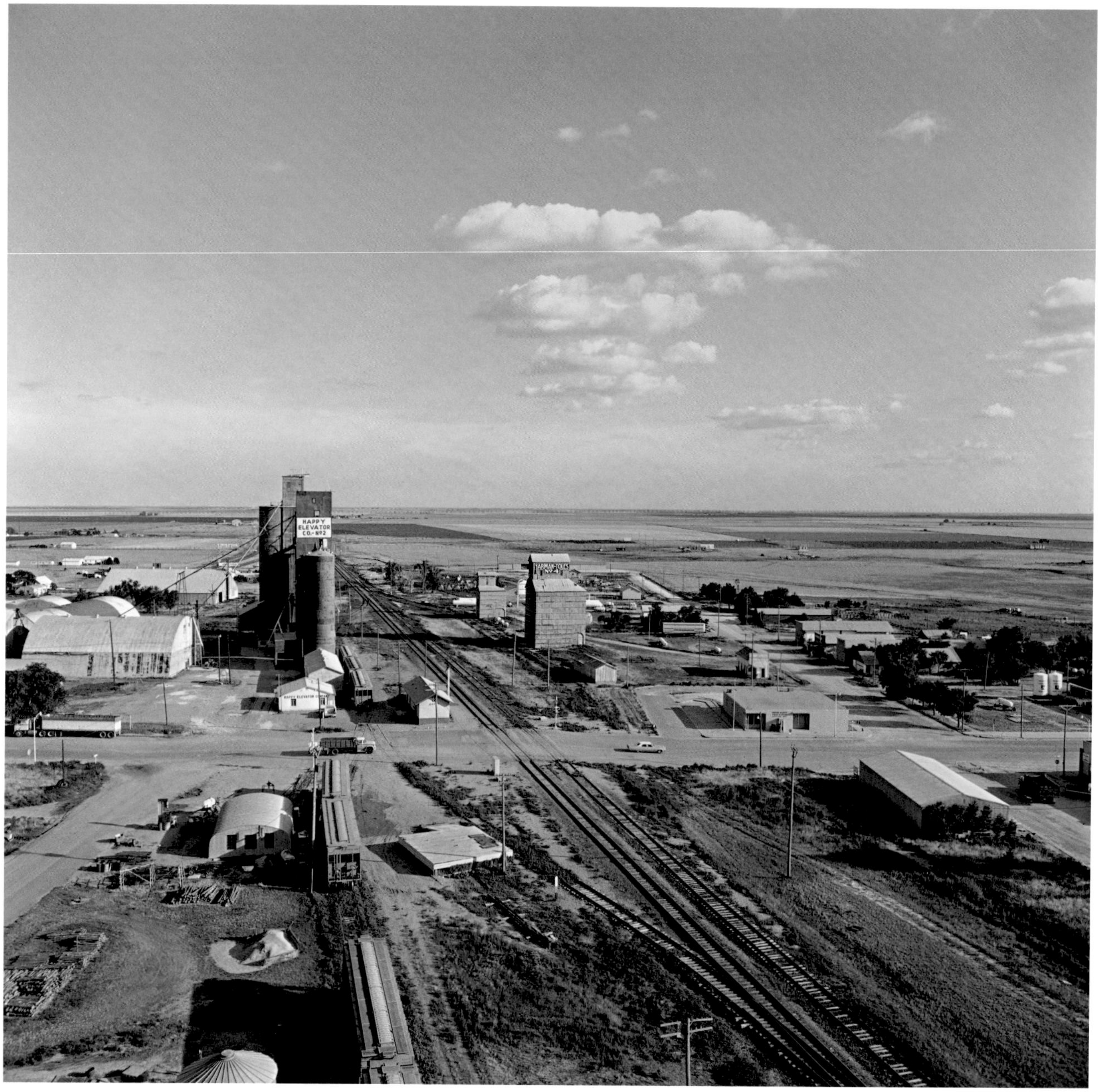

View south from Harmon-Toles elevator / Happy, Texas 1975

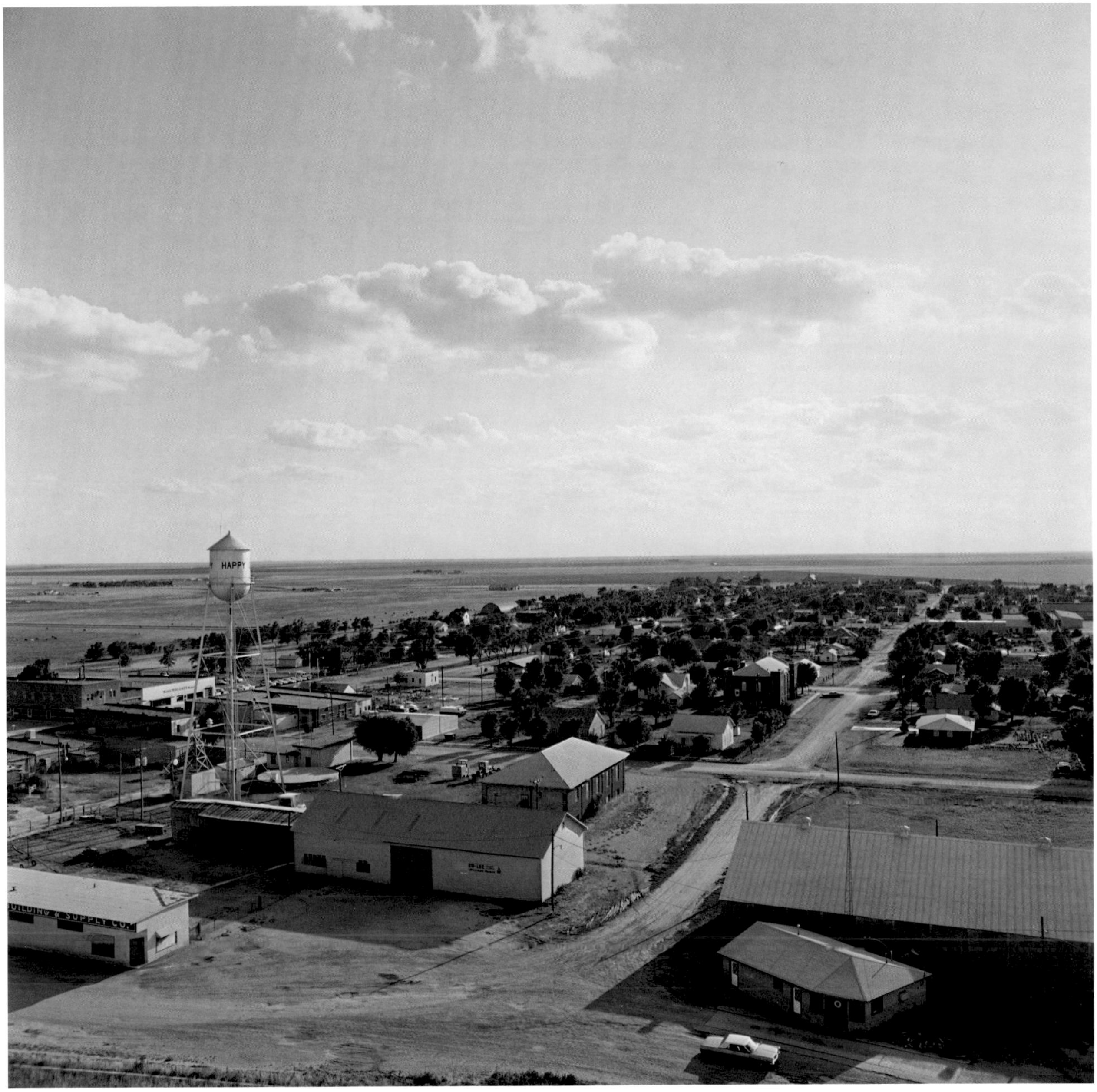

View west from Harmon-Toles elevator / Happy, Texas 1975

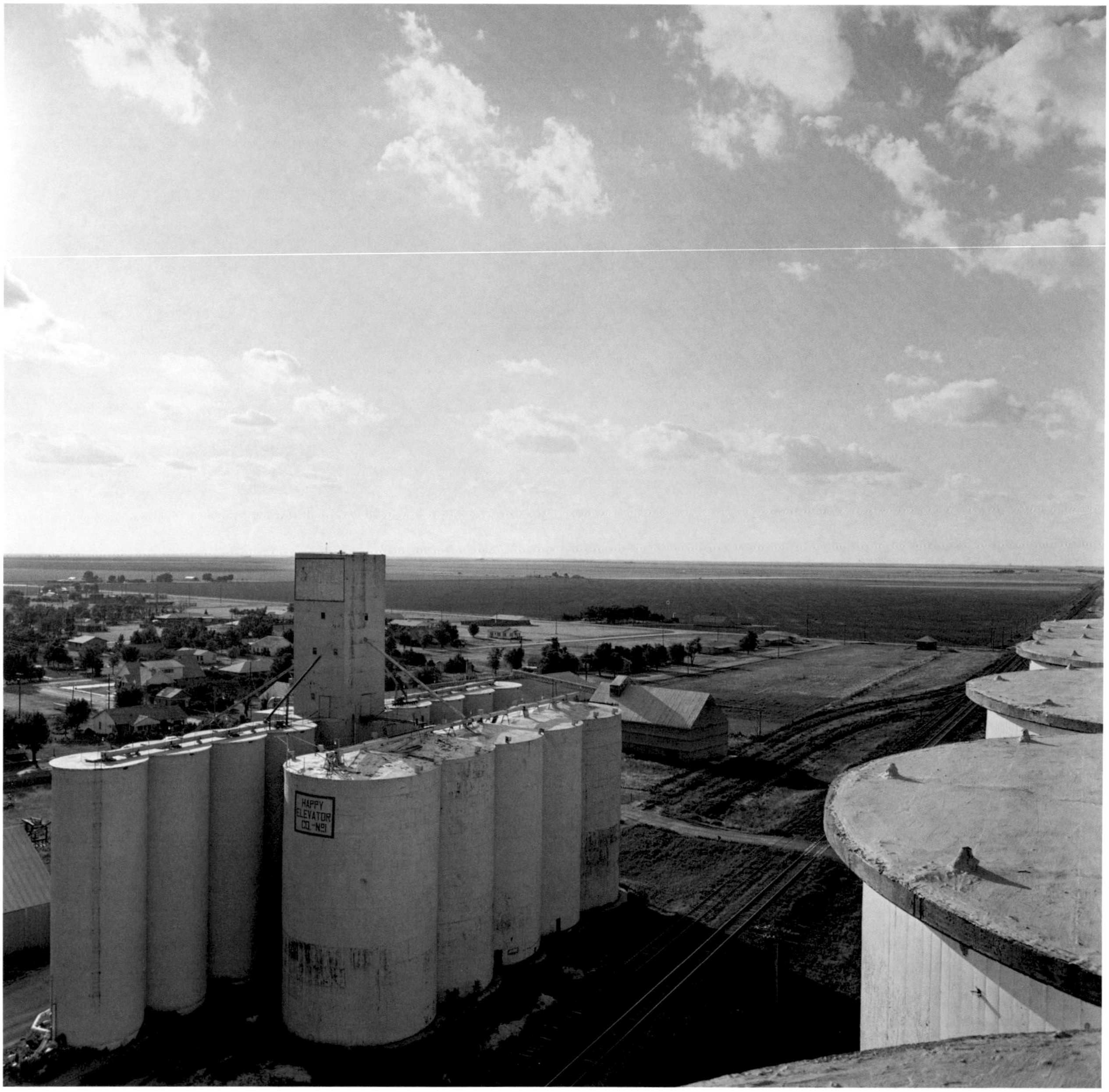

View north from Harmon-Toles elevator / Happy, Texas 1975

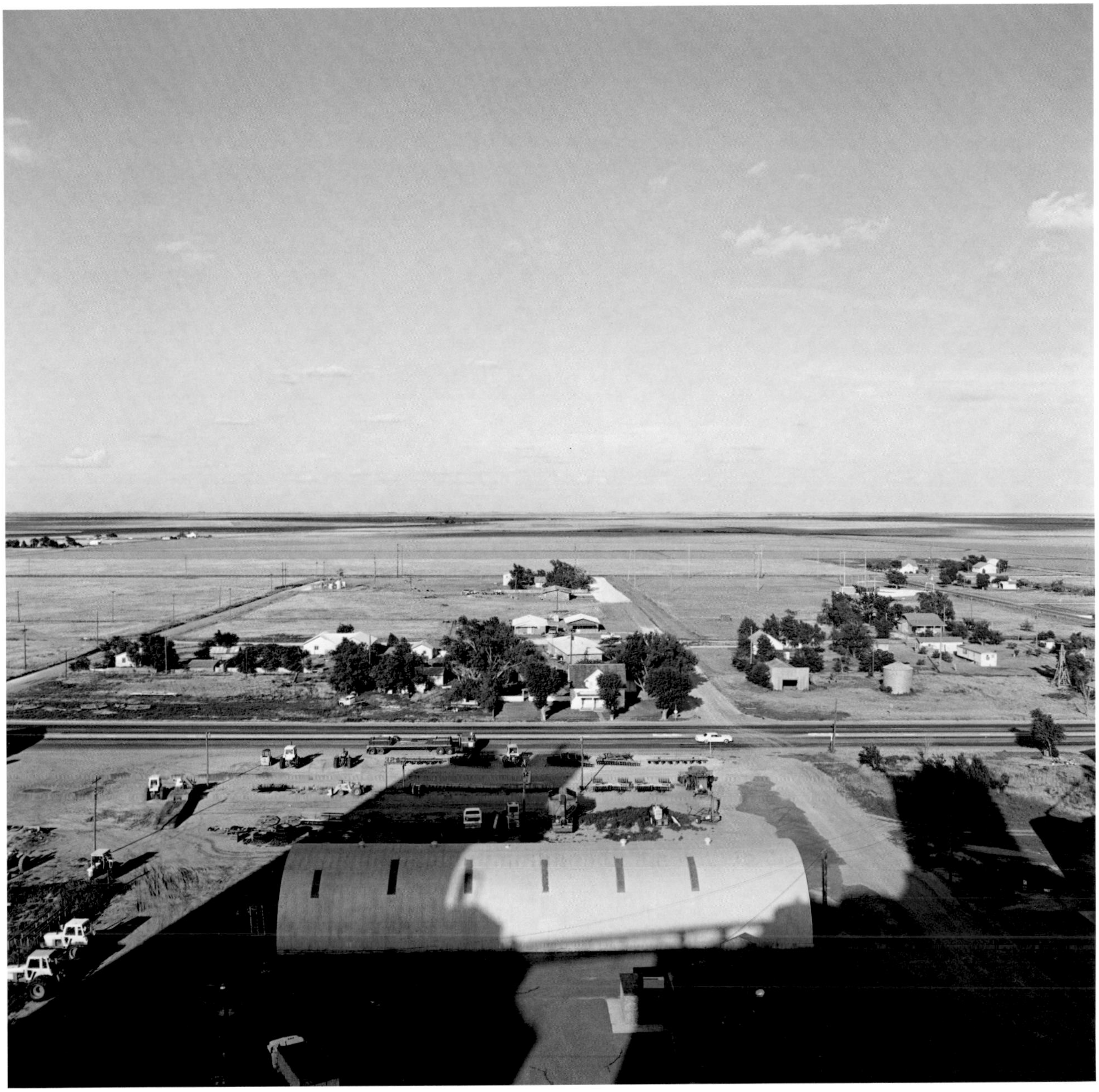

View east from Harmon-Toles elevator / Happy, Texas 1975

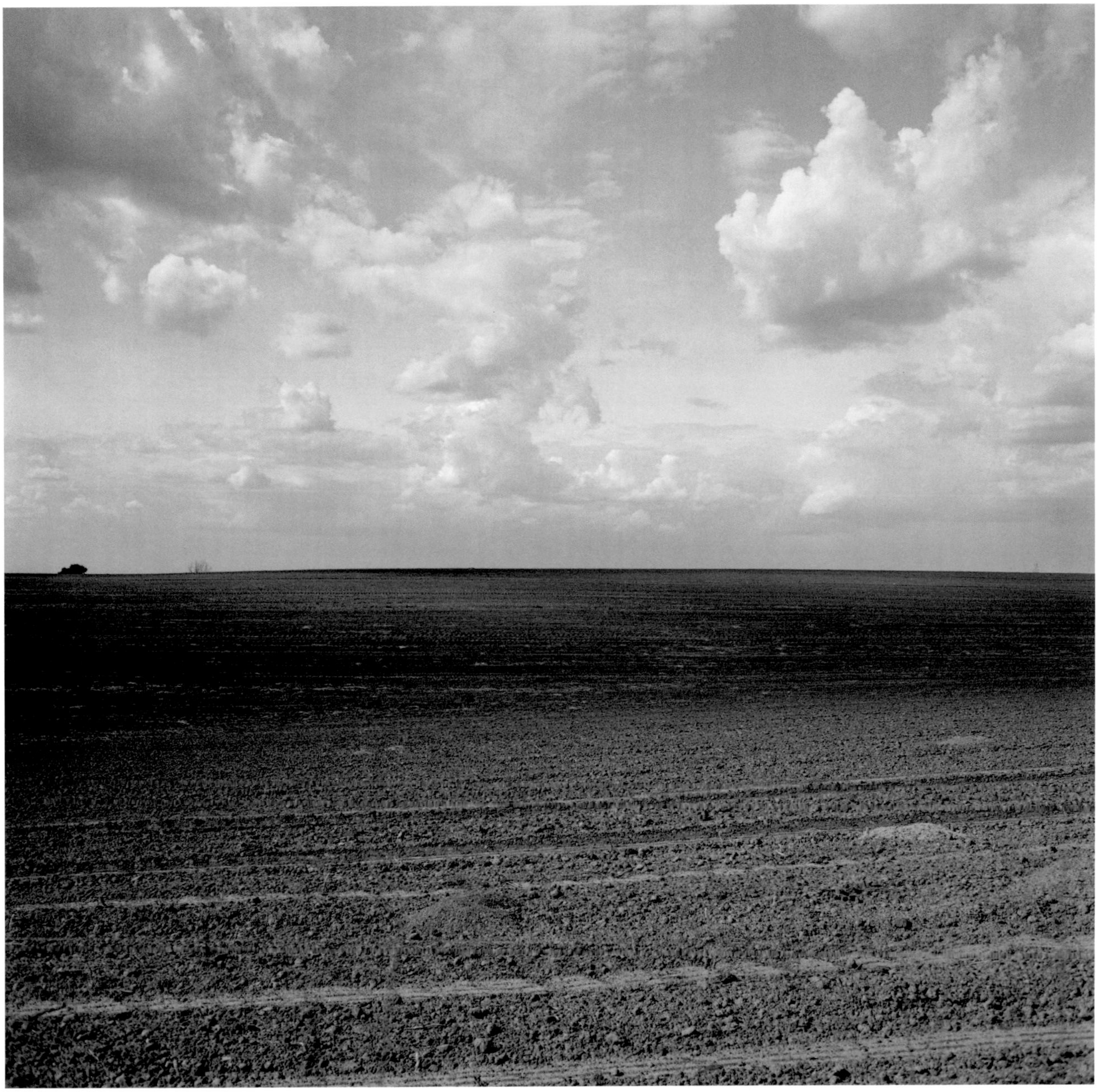

Landscape / near Red Wing, Minnesota 1973

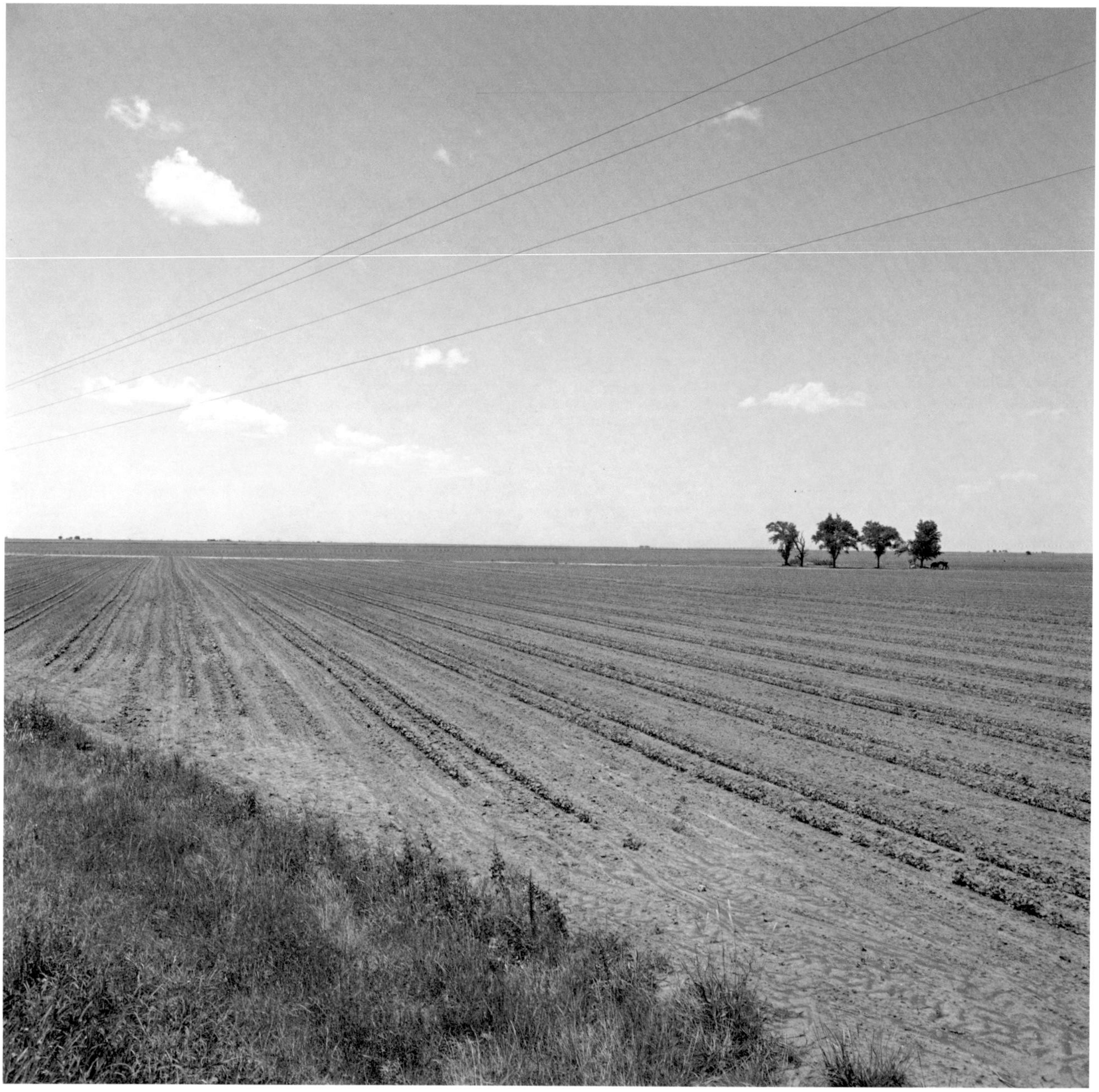

Landscape / near Roscoe, Texas 1975

Landscape / near Littlefield, Texas 1975

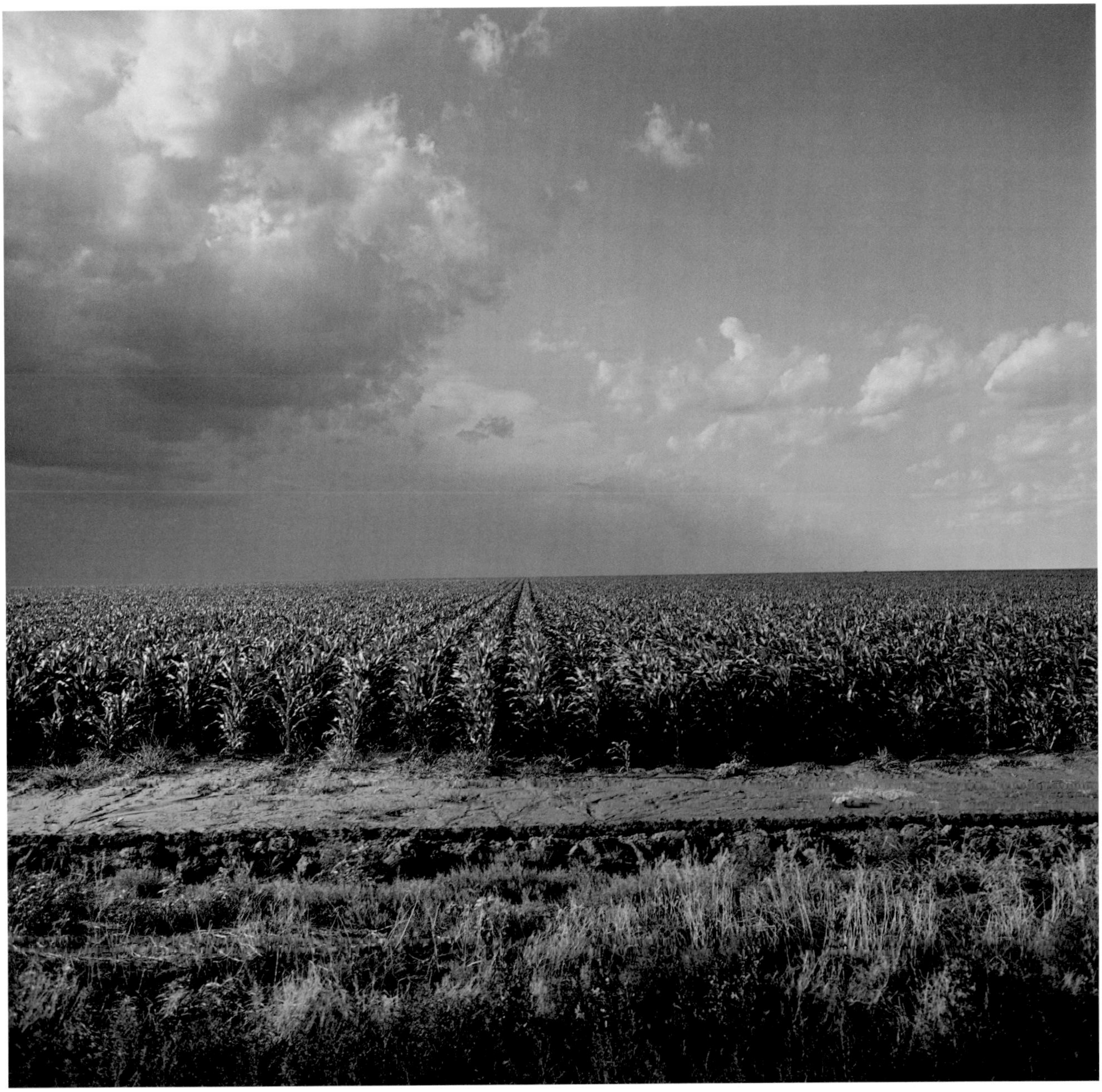

Cornfield and approaching thunderstorm / near Earth, Texas 1975

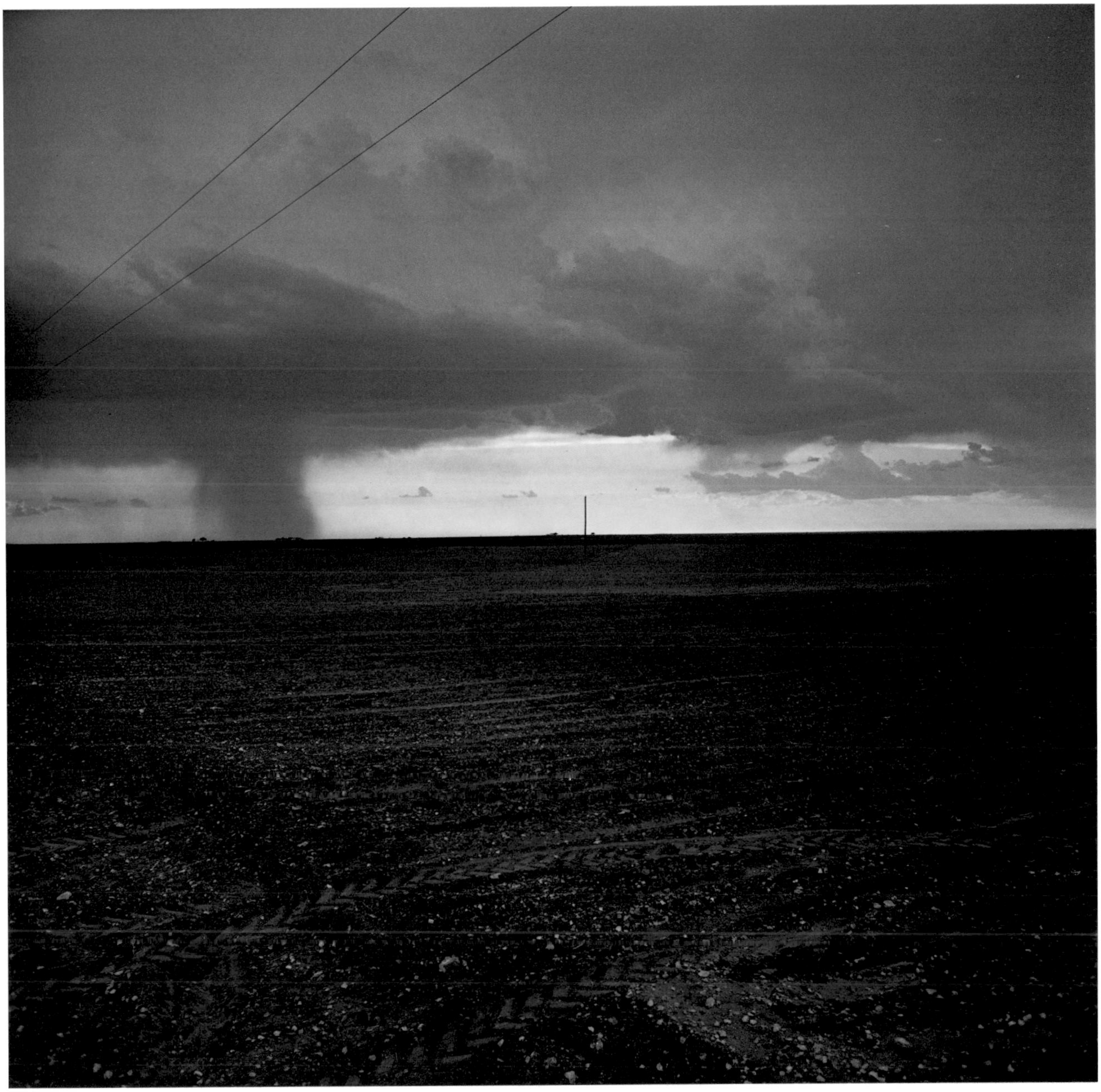

Landscape / near Lamesa, Texas 1975

The Grain Elevator:
An American Invention

The Grain Elevator: An American Invention

by John C. Hudson

They are an American invention, as common to the plains of the United States and Canada as anything save, perhaps, the rolling fields of grain above which they rise. They may store soybeans in Illinois, corn in Iowa, rice in Arkansas, winter wheat in Kansas, durum in North Dakota, barley in Montana, milo in Texas, or spring wheat in Saskatchewan. Almost identical structures can be found at trackside elsewhere, sheltering a crop of hazelnuts in Oregon's Willamette Valley or holding a few hundred tons of catfish meal in the Yazoo Delta of Mississippi. In fact, the grain elevator is at home nearly everywhere in the American agricultural landscape, and its territorial reach continues to grow. Corn and soybean elevators have sprung up where cotton gins once stood in the South, and shiny metal grain bins dwarf abandoned creameries and milk plants along railroad sidings in Wisconsin and Minnesota where a similar shift to cash-grain farming has occurred.

The grain elevator has evolved continuously over the course of American agricultural history. Its function—to store grain between harvest time and shipment to market—has changed scarcely at all, but its appearance has repeatedly been transformed as agriculture has grown in productivity and transportation technology has evolved in response. The basic structure is known as an "elevator" because it typically contains one or more vertical conveyors for lifting grain from ground level to an overhead system of belts and chutes that redirects the grain into various storage bins. From there, it is lofted once again and then fed by gravity or mechanical means into waiting railroad cars, trucks, or barges for shipment to market. Elevating the grain is merely a handy means of transferring it from one bin to another, although grain storage facilities of all kinds are commonly known as elevators whether they contain such an apparatus or not.

Grain merchants classify elevators into three basic types. Country elevators are located along railroad sidetracks in towns or, in some cases, about halfway between towns, and they receive most of their grain from local farmers. Terminal elevators receive their grain via rail or truck from the country elevators, and they sell the product to manufacturers or store the grain for shipment to distant domestic and foreign markets. In recent years a third type, known as a subterminal elevator, has been recognized. Subterminals receive shipments from smaller country elevators for the purpose of assembling large-volume shipments to terminal elevators, thus serving as intermediate storage sites or "surge tanks" in the system.

For many years the typical country elevator on the prairie was a three- or four-story wood-frame structure that had a storage capacity of approximately thirty-five thousand bushels (page 47). The elevator actually handled more grain than that during a typical year because stored grain was shipped out as more was brought in. An elevator might turn over its grain stocks at least four times in a year, creating an actual volume of one hundred fifty thousand bushels or more. If the main crop was wheat, with a yield of twenty-five bushels to the acre, then a single elevator could accommodate more than six thousand acres of the crop. Translated into square miles, this meant that a single elevator could serve an area of perhaps twenty square miles, given that other crops and pastureland were included in the mix of land uses. Translated into the farmers' terms, the travel distance to the elevator via section-line roads would work out to a maximum of about five miles.

The prairie landscape grid of farms, fields, roads, towns, and railroads has a neat, rational geometry about it that makes such calculations simple to perform, yet things were never quite as simple as that for most grain farmers. A five-mile trip to the elevator was really a ten-mile round trip, one that had to be made repeatedly as the harvest was brought in. In the northern plains, where the harvest comes in the fall, the last trips to the elevator sometimes were made in blizzard conditions. Every farmer had to wait his turn in line at the elevator. Sometimes he needed to return to the farm in the wee hours of the morning, only to face the same cycle again the next day. It was not uncommon for bankers to keep their doors open well after midnight during the harvest season to handle the financial transactions that accompanied this once-a-year peak in business.

Virtually all calculations in the grain business are made beyond the first decimal point because the effects of small differences in prices or costs are multiplied thousands of times over. A half-cent difference per bushel in the railroad rate for shipments between one elevator and another translates into a one-hundred-dollar difference for the grain produced on a section (640 acres) of land. A three-quarter-mile difference between a trip to one elevator and another becomes the better part of an extra day's work if you have to travel the same distance, back and forth, forty times. Differences of this magnitude might lead farmers to choose one town rather than another to market their grain. The same margins multiplied by ten sometimes were enough to force a farmer to decide whether he should raise grain at all.

The sheer size and bulk of a grain elevator is magnified by its surroundings, typically a small town on the open prairie. Grain crops such as corn and wheat are produced on the flat to gently rolling lands of the North American prairie because these crops are ideally suited to the environment. They need warm summers, tolerate cold winters, and require only moderate amounts of rainfall. The prairie grasslands, plowed up by the incoming waves of would-be grain farmers a century and more ago, owe their existence not only to the climate but also to fires, which, for thousands of years, systematically eliminated woody plants from the flat upland surfaces. The resulting soils were deep in profile, high in organic matter content, and rich in the nutrients that cereal crops demand. Alternatively, the interior prairies of North America have been used for little else other than grain farming ever since they were settled. Given the comparatively small population required for this activity, it is not surprising that enormous elevators, minuscule towns, and broad expanses of flat land nearly always occur together, forming a natural and human landscape in which each component suits the others.

One of the great myths that shrouds the settlement history of the Great Plains focuses on the railroads as civilizing agents. During the middle to late nineteenth century, when the first transcontinental lines were built, railroads actually did open up new territory to settlement, but outside of these few narrow swaths across the plains from the Dakotas to Texas, railroad lines generally *followed* farmers into new territory, waiting until the course of development was well established before they risked capital in constructing new lines of track. Homesteaders unloaded their belongings at the nearest railhead and trekked across country to stake their claims. When enough of them had arrived to constitute a potential pool of grain shippers, railroads sent field engineers into the territory to collect data on the people and the land they had started to farm and then made cost and revenue projections, much in the manner in which shopping center locations are determined in cities today. New lines of track were extended if enough business was to be had, and construction was hastened if it appeared to one railroad president as though another had designs on the same territory.

A farmer located thirty miles from a railroad generally raised little grain for sale off his farm because he was too handicapped by his remote location. But one located within, say, ten miles of a railroad could be counted on to make grain farming his specialty. Every railroad company knew that a new line of track would be accompanied by a rapid upsurge in grain farming along the route. The farmer's need was for a conveniently located elevator. The railroad's interest, in contrast, was to build as few miles of track as were necessary to gain the largest possible volume of grain traffic. The more productive the countryside, the more miles of track that were built by competing railroads. Railroads hauled people, mail, goods, and livestock as well, but the income from those sources was a minor factor in deciding where and when to build a new line. For this reason, the best grain farming country in the plains was criss-crossed by railroad tracks of two, three, or four competing companies, whereas the marginal areas more suited to livestock ranching saw little railroad construction.

The grain elevator was the link between the farmer and the railroad, and it is little wonder that both sides tried to control the elevator business to suit their own purposes. Railroads had the obvious advantage because an elevator without railroad service was of no use to the farmer. As a further anchor on farmers' trading habits, railroads entered the townsite business by platting trade-center towns adjacent to virtually every location where they decided to locate a sidetrack that would serve an elevator (pages 74 and 75). Some towns along lines of the western railroads were platted by private parties or syndicates of individuals who were primarily interested in real estate sales, but the railroads commonly rebuffed the schemes of outsiders if they conflicted with the company's plans. The grain elevator was the key to the railroad's interest in a town, and for that reason one has to look far and wide to find a railroad town built without one.

Since grain moved from every station along a line to the same milling centers or deepwater ports, it was probably inevitable that some entrepreneur would conceive the idea of a chain of country elevators, one per town, that would funnel traffic more efficiently toward the market. The man generally credited with the idea was a native of the Maine woods named Frank Peavey, who created a system of warehouses along the railroad line linking his terminal elevator in Sioux City, Iowa, with Yankton, South Dakota. In 1875, Peavey convinced officials of the Chicago, St. Paul, Minneapolis & Omaha Railroad to grant him grain elevator locations along all of their lines. Linking the terminal elevator with a line chain of country elevators appealed to men in the milling business and also to railroaders, although railroad corporations as such rarely owned the individual country elevators.

It soon became apparent to railroad officials that the line elevator chains would have to be controlled. The ideal way to do that was to lease elevator sites in every town to rival line chain companies, thereby forcing them to compete with one another, but to do so only if they agreed not to enter into similar agreements with competing railroad companies. When a new line of track was laid, it was common to number the elevator sites (125-foot-wide lots along the tracks) so that line chain *A* got site 1 in every town, line chain *B* got site 2, and so on.

Had the grain business developed differently, it might have transpired that every town would have had a single large elevator. Instead, the result was a row of identical thirty-five-thousand-bushel elevators lined up at trackside, distinguishable from one another only by the names "Monarch," "Imperial," "Occident," "Great Western," or "Peavey" painted high on the sides of the drab red structures. Each elevator competed with the others by shaving its profit margin in order to attract business. It could not divert business to another railroad that might have served the same town without running the risk of losing the elevator site lease. Railroads controlled the elevators, as well as other line chain companies such as lumber yards and fuel dealers, by leasing (and not selling) land along the tracks, although every other parcel of real estate on a railroad townsite was for sale.

Line chain elevator companies, which could command the capital necessary to erect dozens of elevators in a single season, hired contractors who specialized in elevator construction to build on every site the railroad had leased to them along a new line of track. Contractors used the same blueprints over and over, thereby stamping the same jagged silhouette of elevators against the horizon of every town. The better the country was for farming, and the larger the trade area of the town, the more elevators were built. After the construction phase had passed, local managers were hired to transact company business at each site. The farmer was doubly leveraged by this system, which was planned, financed, and built by men he would never see. The railroad had the upper hand in controlling the line chains, but the farmer had to accept terms dictated by both the railroad and the elevator companies.

Continuous fluctuations in prices have long made it necessary for both buyers and sellers of grain to be aware of up-to-the-minute quotations in the terminal markets. Elevator managers in even the remotest locations were linked into the market information network, and every farmer, in turn, had to evaluate the information to decide when, and how much, to sell. A typical transaction that would have taken place at one of the nearly two thousand country elevators in the state of Iowa might be summarized as follows.

Suppose that a farmer wished to sell number 3 yellow corn. He would contact the buyer at his local elevator, who then obtained the latest Chicago market quotation of, perhaps, 60 cents a bushel. A freight rate of 12 cents per hundred pounds from this station made a transportation charge of 6¾ cents based on a weight of 56 pounds per bushel. The terminal market collected 1½ cents' commission for handling the bushel of corn when it arrived at the end of the line. The country elevator normally charged another 3 cents for handling, although this varied depending on how eagerly the farmer's business was sought. The buyer at the elevator therefore would have offered the farmer a price that was between 11¼ and 9¾ cents under the Chicago market quotation.

Whether the farmer had a poor crop or a good one that year, in terms of either yield or grade, was of no consequence in the transaction. The grain collection system was always paid before the farmer was because the transportation and handling costs were subtracted before any money changed hands. The farmer could protect himself against fluctuating prices by buying or selling "futures," agreements to deliver grain at a specified price at a future time. These contracts can be met either by delivering the grain (which generally requires on-farm storage facilities) or by turning over the futures when the price is right. Grain elevator companies have long dealt in futures contracts as a means of hedging against price fluctuations. Farmers with access to continuous market information also deal in futures contracts today, but the practice was uncommon among them before the modern era of agribusiness.

What farmers did seventy-five or one hundred years ago, instead, was to organize themselves into marketing cooperatives to avoid dealing with the line chain companies altogether. Deals made in the grain business had already led to consolidation of ownership over the line chains as the smaller chains sold out to larger ones or unprofitable lines were sold to local parties. Farmers in a locality found it relatively easy to purchase one or more country elevators from the line companies. Under the leadership of major marketing cooperatives (such as the Farmers Union Grain Terminal Association in the Dakotas and Montana or the Union Equity Cooperative Exchange in Kansas, Oklahoma, and Texas), farmer-owned elevators eventually took on all the business operations of a line company, including ownership of some of the terminal elevators. The new system, which emerged in the first decades of the twentieth century, allowed farmers to share the profits, which the lines had heretofore taken for themselves.

The elevators began to take on a new look as well. Old corporate identifications were painted over, and names like "Farmers Union Elevator," "Equity Grain Association," "Producers Grain Coop," or, in Canada, "United Grain Growers" replaced them. New five-thousand-bushel grain bins were erected next to the old elevators. Metal spouts replaced the wooden troughs used to funnel grain into waiting railroad cars. Improved weighing and cleaning facilities were installed in the elevator cupolas, and cyclone dust collectors, designed to prevent grain dust explosions, began sprouting from elevator sides (pages 32 and 33). No longer the private property of an alien corporation, the elevator increasingly became a friendly gathering point for the farmers who owned it. The elevator was the farmers' piece of a town's turf, all but off-limits to town residents, many of whom probably did not realize that it was a place literally hard-wired to the world outside, where a political assassination in Europe or news of a crop failure in Argentina was instantly received in the form of ¼-cent increments or decrements in the price of grain.

The production, marketing, and manufacture of grain—including corn, wheat, rice, and soybeans—is the world's largest industry, and it has held that distinction for many years. The interior plains of the United States and Canada, in turn, have long been the most important grain-producing region in the world. The generic term *grain* masks important differences between various crops in terms of marketing and distribution, however. Corn, as well as its relatives such as milo maize, is used mainly for fattening meat animals. Only about one-fourth of the American corn crop is exported. Soybeans, which began appearing in rotation with corn on midwestern farms in the 1940s, are crushed for edible oils and other by-products, and about 40 percent of the crop is exported. On the other hand, more than 60 percent of the American wheat crop has been sold abroad in some recent years.

Corn moves through local elevators on its way to distant feedlots or corn oil processors, although a substantial amount has traditionally been consumed on the farms where it is produced. Nearly all of the wheat, rice, barley, and soybean crops are handled through local elevators and from there are sent either to domestic processors or to ocean ports. The largest concentrations of grain elevators per square mile are found in the wheat- and soybean-producing areas of the country, correspondingly fewer in corn country. The changing balance between domestic consumption and foreign sale of grain has not affected the make-up of the country elevator to any great extent, but it has wrought considerable changes in the role of terminal markets.

Increases in agricultural productivity have affected all of these crops. The mechanization revolution of the latter nineteenth century was followed by the internal combustion revolution in the early decades of the twentieth. The introduction of high-yielding varieties and the increased use of fertilizers, pesticides, and herbicides created another jump in productivity per acre in the post–World War II period. The same amount of land, operated by one-fourth as many farmers, may be five times more productive today than it was a century ago. One result, of course, is that grain elevators have had to expand in size to keep pace with this increased output, since the number of elevator sites was virtually fixed more than a century ago when railroads dictated their locations.

The breadbasket of the United States moved west with the frontier, from Connecticut to New York to Ohio, then on to Wisconsin and Minnesota, and finally to the Dakotas, Kansas, and the rest of the Great Plains. The wheat that farmers raised in the early days of the American republic was hauled only a few miles to the nearest water-powered mill where it was ground into flour, as it had been in Europe for centuries. Grain was handled in sacks, hauled in wagons, stacked in flat warehouses, and then wheeled aboard canal boats or coastal vessels for distribution to market. Only one bushel of wheat in every eight produced in the United States in 1820 was raised west of Pittsburgh or Buffalo. The market for the trans-Appalachian West, even through the 1830s, was New Orleans, where grain was reloaded aboard sailing vessels for distribution to the East after it came down the Mississippi on flatboats.

By 1840, nearly one-third of the American wheat crop was raised in Ohio, Indiana, Illinois, Michigan, and Wisconsin. The Great Lakes and the Erie Canal were the routes by which the product moved east, and cities such as Cleveland and Toledo were the major shipping points. Grain handling was still a fairly primitive affair, with many small warehouses assembling shipments. New wheat frontiers in Wisconsin opened in the 1840s, and by the end of that decade more than twenty million bushels were being sent east via Great Lakes vessels. Nearly all of it was destined for the port cities of Buffalo and Oswego, New York, where the first large terminal elevators in the American grain trade appeared.

Continued expansion on the western frontier made Chicago the principal assembly point for grain by 1850. The Chicago Board of Trade, a nonprofit organization with members representing millers, processors, importers, exporters, and elevator interests, was organized in 1848. In that same year, Chicago acquired its first steam-powered grain elevator. The concentration of wheat and corn marketing at Chicago developed rapidly in the next few years, and by 1860 the city had several terminal elevators with million-bushel capacities. Chicago's emerging role in the grain trade depended partly on its parallel role as a railroad center. The tentacles of iron rails that reached from Chicago outward in all directions across the Midwest foreshadowed similar developments that would occur around Minneapolis and Kansas City as the grain frontier moved still farther to the west.

The country elevators that began to appear after the mid-1870s were really little more than scaled-down versions of the early terminal elevators. The basic unit, a bushel of wheat, remained the standard, but no longer was it feasible to handle grain in bushels or even in hundred-pound sacks. Railway technology, however, made a somewhat inadequate response to the new demands placed upon the system. Instead of handling grain in easily loaded and unloaded bulk cars, in the manner, for example, in which coal was transported, railroads chose instead to manufacture thousands upon thousands of all-purpose boxcars. These forty-foot wooden carriages had large doors on both sides that opened from floor to roof, making it easy to load cargo such as boxed manufactured goods, but requiring some modification to be useful for bulk commodities.

For the better part of the next century, nearly all grain handled from country elevators was shipped in railroad boxcars. Boards and paper were placed across the open-door side in order to make it possible to load the car to its fourteen-hundred-bushel capacity without having grain spill onto the ground. When the harvest came, railroads dispatched strings of these boxcars along every line, five or ten per elevator, every few days. The loaded cars were assembled into strings as long as a locomotive could pull, hauled to the nearest railroad terminal, and then reassembled into still longer strings pulled by larger locomotives for the journey to market.

The bulking of grain into ever-larger quantities really began on the farm, where the grain wagons were loaded. A quarter-section (160 acres) of wheat would fill twenty wagons. The grain from each wagon that was brought in to the elevator was graded, weighed, and then redirected through the elevator's maze of bucketed belts and chutes into a bin that contained grain of the same grade. Two dozen wagonloads might fill one bin, which, in turn, provided about enough grain to fill two boxcars. At times, when grain was arriving faster than the elevator could handle it, grain buyers operated "on track," purchasing whatever arrived and shoveling it straight from the farmer's wagon into a waiting boxcar.

Elevator employees worked around the clock in the height of the season. Using a team of horses, they would roll an empty car into place opposite the loading spouts as soon as the previous one was full. Two or three men with grain scoops worked furiously in the heat and dust, evening out the pile on the car's floor as it accumulated, pyramid-like, beneath the grain spout. When the car was filled, the door was sealed, and a consignment label prepared by the local railway agent was affixed to the side indicating the car's destination. "Duluth" or "Superior" meant shipment to a terminal elevator on the Lake Superior docks from which the grain might move to Buffalo or Liverpool or Antwerp. "Minneapolis" cars were most likely destined for a miller in that city, as were those bound for Omaha or Kansas City. "Waterloo," "St. Joseph," or "St. Louis" shipments might end up as livestock feed.

The terminal elevators where these shipments arrived were not terminals in the sense of a final destination. There, the same steps that had already been performed at the country elevator were repeated. Cars were unloaded by shoveling the grain out through the boxcar doors into a trackside grate or, in later years, by spotting the car on a rotating carriage that tipped it back and forth to allow the car's contents to empty. Stored grain was later reloaded aboard ships or else poured back into more empty boxcars for shipment to millers and other grain processors. Cumbersome though these operations were, they were profitable if undertaken in large enough volume.

Two visual images of the American grain elevator stand in sharp contrast. One is of the weatherbeaten country elevator standing above the prairie. The other is of a massive block of concrete silos that make up the terminal elevator located alongside a railyard or wharf in a city. The country elevator evokes the sense of empty loneliness on the plains; the terminal elevator, in contrast, validates our sense of American agricultural productivity. Probably the elevators most familiar to Americans are two of the largest—the Harvest States Cooperatives complex on the docks at Superior, Wisconsin, and, still larger, Union Equity Cooperative Exchange Elevator "J" at Hutchinson, Kansas, with its eighteen-million-bushel storage capacity. Relatively few people know these buildings firsthand, but photographs of them are often found in school textbooks or in photo-essays on the "American breadbasket."

Other terminal elevator cities, such as Thunder Bay, Ontario, Enid, Oklahoma, or Council Bluffs, Iowa, are also an integral part of the system. It is difficult to identify a typical terminal elevator organization, but the Omaha Grain Exchange, representing a dozen terminal elevators totaling more than thirty-two million bushels' storage capacity plus four flouring mills in Omaha, Council Bluffs, and nearby Ralston, Nebraska, can serve as one example. The Omaha area receives in excess of one hundred million bushels of corn, wheat, and other grains annually. About three-fourths of the grain arrives by truck, chiefly from western Iowa and southeastern Nebraska, which makes an average of two hundred seventy-five loaded grain trucks entering the Omaha area each day from farms and country elevators. Another six thousand loaded rail cars arrive at the Omaha–Council Bluffs terminal elevators every year.

Part of the incoming quantity is sold locally to millers, but most of it eventually moves out of the Omaha area. Under the transportation rate structure now in effect, rail shipment is generally the most economic alternative, and nearly all of Omaha–Council Bluffs' outmovements are handled in that manner. Twenty years ago, when barge rates were more competitive on the Missouri River, about one-fifth of the outbound grain was handled on barges, but railroad rate reductions cut into the barge business and, even more significantly, into the grain trucking business. Fraction-of-a-cent differences in per bushel costs are again evident, in this case enough to produce switches between modes of bulk transportation.

In the 1920s, the Omaha Grain Exchange received only two-thirds the incoming grain it gets today, but Omaha mills processed more of the arriving grain then than they do now, reflecting a trend toward greater concentration in the milling industry. Flour milling once was a highly localized, small-scale industry that could be carried on along any stream capable of turning a water wheel. When the United States was a nation of farmers, the manufacture of flour involved little but grinding what a farmer brought to the mill and returning it to him. A few larger mills manufactured flour for urban markets, but they, too, were commonly located near the point of consumption. This system worked well when the population frontier had yet to reach the Great Lakes, but it disappeared rapidly as soon as grain production was established on a large scale west of Chicago. Flour milling was drawn to intermediate locations, between the wheat-producing areas and the market. Milling cities thus became the typical locations for terminal elevators.

In the past century, the commercial baking industry has grown to the point that home flour consumption has nearly disappeared. The upgrading of flour standards, evolution of brand and grade preferences, and economies of scale in the industry have produced a tendency toward fewer, but larger, firms. Today, about two dozen millers manufacture nearly all of the wheat flour sold in the United States. Most mills still are located within easy access of large terminal elevator complexes, just as they were a century ago. Increased concentration in the milling industry has taken place within the system linking country elevators with terminals, which is not surprising given the dominant role the millers had in organizing the elevator system in the first place.

One of the early improvements in grain storage buildings, which permitted the construction of large terminal elevators, was the introduction, around 1900, of the "slip-form" method of concrete construction, whereby a grain silo could be built continuously by slipping the metal casting forms upward to rest on the previously cast portion. Limits on the height of these cylindrical silos, which had three or four times the capacity of the older wooden structures, were set more by problems of heat and pressure build-up on the grain than by height limitations alone. Eventually, procedures were implemented for regular transfer of grain from bin to bin to retard spoilage and prevent explosions. (The weight of the material generates the high temperatures, high enough to make grain dust into a volatile fuel.) Temperature and humidity sensors were installed at regular intervals inside a silo to monitor conditions. Despite the safety precautions, grain elevator explosions still occur, sometimes with tragic loss of life.

The old thirty-five-thousand-bushel wood-frame elevators were outmoded once the new concrete silo technology was adopted. Many of the old buildings remain today, but they are too small to handle the grain they receive. The cluster of metal grain bins that commonly surrounds an older, wooden elevator is linked to it by spouts tied in to the older elevator's grain-handling machinery. During the 1950s, the federal government made generous loans to farmers' cooperatives to construct new storage facilities. It was a time when many small towns saw a doubling or tripling of local elevator capacity (page 51), a development made necessary by the increased productivity per acre that had come from mechanization and agrochemical innovations.

The first of these portable grain bins appeared under the auspices of the Commodity Credit Corporation (CCC) in 1938; they were used to store grain against default on government loans to farmers. The CCC kept expanding its operations, reaching a peak of 990 million bushels in 1957. A cluster of identical cylindrical, rib-sided, conical-roofed metal bins became a familiar sight in farm country by the 1950s, one that reminded many of the vast overproduction of American agriculture and the cost to the public of storing the surplus. The CCC sold the bins to farmers in the 1970s, by which time the boom in American grain exports was well under way.

Fifty years ago the United States was purchasing wheat from Canada because domestic production could not equal demand; today the nation accounts for approximately one-third of the world's wheat exports. Fifty years ago American farmers produced the world's largest corn crop, fed most of it to animals on their own farms, and exported barely one percent to other countries; today the United States supplies two-thirds of the world's corn exports. Fifty years ago soybeans were a rarity on American farms; today, more than two billion bushels of beans are produced annually, and exports from this country satisfy two-thirds of the international demand.

The geography of domestic grain consumption has changed just as dramatically in the same period. Production of meat animals was once a function performed largely by farms in the Midwest, but the growth of population in the South and the Far West in recent decades was soon followed by migration of the meat-packing industry. Poultry, which looms ever larger in the American diet, is largely produced in the South, and southern farmers also raise beef and pork in far larger quantities than ever before. Southern meat animal producers consume more grain than they produce, satisfying the balance by purchases from the same midwestern farmers who once fed that grain to their own animals. The typical Corn Belt farmer today is a cash-grain specialist whose primary income is received from the sale of crops rather than livestock.

Early in the 1960s, officials of the Southern Railway ordered the construction of a new type of grain car. It resembled a coal hopper car, with dump-doors on the underside but with a covered top, which would protect commodities such as grain from the elements while in transit. Known as "Big John," after a popular song of the day, the new covered hopper car could carry four thousand bushels of grain, nearly three times the capacity of the standard boxcar. It could be loaded from any country elevator and dumped at any terminal elevator, but far more efficiently and without the necessity for massive amounts of hand-shoveling. It was hailed as a technological breakthrough, although it was a long-overdue response from railroad companies, which were fast losing their grain business to truckers and barge lines.

The Southern Railway posted a rate on five-car "Big John" movements that was approximately one-half the rate per bushel for a standard single-boxcar shipment. The Tennessee Valley Authority, representing barge interests, took the railroad to court complaining that a below-cost rate had been set in order to put the barges out of business. Other litigants soon entered the controversy. The case dragged on for four years, was heard thirteen times in the appellate courts and twice before the United States Supreme Court, and was finally decided in the railroad's favor in 1966. Soon, southern farmers were importing midwestern grain at half of what they once had paid.

Within a decade, the four-thousand-bushel covered hopper car became the standard in the industry (page 42). The five-car unit rate originated by the Southern Railway soon was overshadowed by still lower rates on twenty-five, fifty, and even one-hundred-car unit-train shipments. The grain export boom of the 1970s was thus substantially assisted by volume-pricing strategies of the railroads who, for their part, regained millions of dollars of business annually that they had lost to trucks and barges.

As would be expected, the old thirty-five-thousand-bushel country elevator fared badly under the new regime. Too small to take advantage of multiple-car rates, it became increasingly redundant. Railroad branch lines that served only elevators of this size were the subject of abandonment proceedings by the mid-1970s (page 48). By the mid-1980s, the state of Iowa had only one-half the railroad mileage it had in 1920, and other midwestern states saw similar reductions. By this time, plenty of Iowa farmers had on-farm storage capacities greater than the old elevators in town. They hired local truckers to load grain on the farm and haul it to the Mississippi River, where it could be poured into barges for shipment to the Baton Rouge–New Orleans district from which it went overseas. Kansas and Oklahoma wheat farmers had such an outlet at Tulsa (Catoosa) on the Arkansas Waterway, while Montana barley and wheat growers could take advantage of the Columbia–Snake waterway from Lewiston, Idaho, to Portland, Oregon. Rail, truck, and barge were thus put into growing competition for an increasingly lucrative business.

The competition was largely responsible for the creation of a new type of grain elevator known as the subterminal. An obvious cross between the old country elevator and the large terminal elevator, the subterminal is some distance from the terminal market, but its storage capacity, and therefore its outward appearance, is closer to that of a small terminal elevator. Most of the new subterminals are at access points convenient for assembling shipments via both rail and highway. Most grain arrives at the subterminals via truck—or even tractor-pulled grain wagons straight from the farm—and is shipped out in four-thousand-bushel railroad grain cars in unit-train movements. Often as not, the train transfers the grain to a barge terminal along one of the major rivers. All the modes of transportation are thus combined to achieve the lowest possible rates.

Located in open country away from a town, or tucked in a remote corner of a railyard next to a freeway overpass, subterminals are not inspiring structures. Grain comes in from someplace and is shipped out to some other place, but none of it has much to do with the locality. The long lines of horse-drawn grain wagons at the country elevator of yesteryear have been replaced by long lines of semitrailer trucks waiting to unload at a half-million-bushel subterminal. The queue is the same because the input-loads and the storage capacity stand in about the same ratio; only the volume has been multiplied, ten times or more. Sleepy-eyed truckers use the delay to get some rest before returning with empty vehicles to the country elevators they are hauling from. When the price is right, grain moves, and those who do the moving get little rest, as always.

The new subterminals are useful storage sites in the grain pipeline. The American milling industry works on a thirty- to forty-five-day inventory of raw materials, but grain of a given variety is harvested only once a year in the various regions of production. The massive terminal elevators, concentrated in only a few dozen cities, are less conveniently located than the more widespread network of subterminals. Grain that stays "out in the country" in a subterminal is subjected to less of the cross-hauling that used to take place when a sudden shift in demand occurred.

The subterminal boom of the late 1970s and early 1980s added more elevator capacity to the system. Some country elevators were put out of business, but many others have remained open to serve as satellites for the subterminals. The result is an excess of capacity. For example, the entire wheat crop of Kansas this year could be stored in the state's elevators with enough room left over for all the wheat produced in North Dakota as well. There is excess capacity in the system because of the empty country elevators now bypassed as storage sites. Agricultural economists in South Dakota have projected that 60 percent of the operating grain elevators in that state are now redundant. Their capacities are too small, their locations along weed-overgrown railroad branch lines are too remote, and their handling facilities are too antiquated.

What will happen to the many small towns whose only remaining economic function is the elevator? The vision of widely scattered subterminals, perhaps one huge elevator serving two counties, is held by some. In recent years, however, other country elevators have received new life as a result of railroad reorganization involving sale of branch lines to local groups of grain shippers. "Spin-off" railroad short lines can handle grain (the only reason for the line's existence) more cheaply than larger railroad companies, which have costly agreements with organized labor. Economic efficiency and dealing in volume have always characterized the grain business. Today, no less than in the past, the least-cost solution is likely to prevail. Industrial deconcentration lowers costs enough to make nineteenth-century technology competitive once again.

Thousands of small prairie towns created by railroad companies a century ago have been reduced to little more than grain marketing points because economic changes have made a dense network of small trade centers obsolete. But the railroads that dictated the very existence of these towns never were interested in them as communities of people. Towns were, in the railroads' view, simply locations where grain elevators were built. Banks, grocery stores, implement dealers, schools, and the rest of what constituted a town were necessary in order to attract the farmer to that point for trading purposes in the early years, but the loss of practically all of these functions in the past four decades did not diminish the town's usefulness to the railroad as long as the elevator remained. Today, of course, there is no particular need for grain elevators to be in towns at all, given the mobility of farmers. The massive concrete silos of the modern subterminal elevator may have dealt the last blow to the viability of many small towns.

In the midsummer of 1991, grain elevators in the Midwest were working around the clock, pouring millions of bushels from storage into railroad grain cars. The boom was created by a sudden upsurge in demand, this time the result of large grain purchases by the Chinese. Blocks of cars assembled in the heartland became a parade of one-hundred-car-unit grain trains creaking and groaning south through Fort Worth or Memphis on their way to Gulf Coast ports. The latest upsurge will soon be only a memory, but it will eventually be replaced by yet another. The elevators that assemble each new shipment, be it destined for Cedar Rapids or Shanghai, will continue to perform the function they have served for more than a century, linking the grain farmer and the market via whatever transportation is cheapest, storing what has been produced until it is ready for delivery. Their stark forms on the landscape will change in response to new technology, just as they did in the past; but whatever shape they assume, the grain elevators of the American prairie will stand as visual reminders of calculated efficiency in an intensely competitive business.

About the Authors

Frank Gohlke was born in Wichita Falls, Texas, in 1942. He attended Davidson College in North Carolina and earned his B.A. in English at the University of Texas in Austin and an M.A. in English literature at Yale University. While living in Connecticut, he studied photography with Paul Caponigro and was also encouraged by Walker Evans to make photography his lifework. After three years in Vermont, he moved to Minneapolis, where he lived and worked for the next sixteen years. During that time he received two National Endowment for the Arts Fellowships, a Bush Foundation Artists' Fellowship, and two John Simon Guggenheim Memorial Foundation Fellowships. In 1987 he moved to Massachusetts. Frank Gohlke's work has been exhibited widely in the United States, Europe, and Japan; he has been given one-person shows at the Museum of Modern Art in New York City, the Art Institute of Chicago, the George Eastman House in Rochester, New York, the Amon Carter Museum of Western Art in Fort Worth, Texas, and the Museum of Contemporary Photography in Chicago. A fifteen-year survey of his work, *Landscapes from the Middle of the World,* was published in 1988 by the Friends of Photography and the Museum of Contemporary Photography.

John C. Hudson was born in Milton, Wisconsin, in 1941. He was educated at the University of Wisconsin / Madison (B.S., geography) and the University of Iowa (M.A. and Ph.D., geography). He has been a John Simon Guggenheim Memorial Foundation Fellow, and he served as editor of the *Annals of the Association of American Geographers* from 1976 through 1981. He is the author of *Geographical Diffusion Theory* (1972) and *Plains Country Towns* (1985), which was awarded the John Brinckerhoff Jackson Prize. He is professor of geography at Northwestern University in Evanston, Illinois.

Books in the Series

The American Backwoods Frontier: An Ethnic and Ecological Interpretation
Terry G. Jordan and Matti Kaups

The City Beautiful Movement
William H. Wilson

The Rough Road to Renaissance: Urban Revitalization in America, 1940–1985
Jon C. Teaford

Greenways for America
Charles E. Little

Bravo 20: The Bombing of the American West
Richard Misrach, with Myriam Weisang Misrach

The Spanish-American Homeland: Four Centuries in New Mexico's Rio Arriba
Alvar W. Carlson

Nuclear Landscapes
Peter Goin

The Last Great Necessity: Cemeteries in American History
David Charles Sloane

To Build in a New Land: Ethnic Landscapes in North America
Edited by Allen G. Noble

Measure of Emptiness: Grain Elevators in the American Landscape
Frank Gohlke, with John C. Hudson

Jens Jensen: Maker of Natural Parks and Gardens
Robert E. Grese

Copyedited by Carol Ehrlich
Designed by Lorraine Ferguson
Duotone separations by Robert Hennessey
Printed by Litho Specialties, Saint Paul, Minnesota
on 100-lb. Lustro Offset Enamel Dull Text
Bound by Midwest Editions, Minneapolis, Minnesota